Modern Views
of Electricity

OLIVER LODGE

CAMBRIDGE
UNIVERSITY PRESS

CAMBRIDGE UNIVERSITY PRESS

Cambridge, New York, Melbourne, Madrid, Cape Town,
Singapore, São Paolo, Delhi, Mexico City

Published in the United States of America by Cambridge University Press, New York

www.cambridge.org
Information on this title: www.cambridge.org/9781108052177

© in this compilation Cambridge University Press 2012

This edition first published 1889
This digitally printed version 2012

ISBN 978-1-108-05217-7 Paperback

MODERN VIEWS

OF

ELECTRICITY

NATURE SERIES

MODERN VIEWS

OF

ELECTRICITY

BY

OLIVER J. LODGE, D.Sc., LL.D., F.R.S.

Professor of Experimental Physics in University College, Liverpool

WITH ILLUSTRATIONS

London

MACMILLAN AND CO.

AND NEW YORK

1889

Richard Clay and Sons, Limited
London and Bungay.

CAMBRIDGE LIBRARY COLLECTION

Books of enduring scholarly value

Technology

The focus of this series is engineering, broadly construed. It covers technological innovation from a range of periods and cultures, but centres on the
technological achievements of the industrial era in the West, particularly
in the nineteenth century, as understood by their contemporaries.
Infrastructure is one major focus, covering the building of railways and
canals, bridges and tunnels, land drainage, the laying of submarine cables,
and the construction of docks and lighthouses. Other key topics include
developments in industrial and manufacturing fields such as mining
technology, the production of iron and steel, the use of steam power, and
chemical processes such as photography and textile dyes.

Modern Views of Electricity

In 1889, a year after both he and Heinrich Hertz discovered electromagnetic
waves and for the first time demonstrated the truth of Maxwell's great
theory of the electromagnetic field, physicist Oliver Lodge (1851–1940)
published his deepest reflections on the nature and meaning of electricity,
how it originates, and its different manifestations. There had been great
scientific advances – the work of Faraday and Maxwell, his own experiments
and those of Hertz – and a revolution in technology. There were also
puzzling questions. What is the connection between electricity and the ether
that occupies space? How does electricity manifest itself in matter? Why
does it come in fixed units? The discovery of the electron eight years later
would offer crucial answers. Always lucid and direct, with a gift for making
the difficult seem simple, Lodge engages the reader with his fascination for
the subject, much as he did in his famous lectures.

ADVERTISEMENT.

THE object of this work is to explain without technicalities, and to illustrate as far as possible by mechanical models and analogies, the position of thinkers on electrical subjects at the present time. It deals particularly with that view of electrical theory which is specially associated with the names of Clerk-Maxwell and Sir William Thomson, and it aims at going as far as possible into the most recondite portion of the subject, explaining what is known of the nature of electricity, but entirely without the use of mathematics. The subject is divided by the author into four parts: (1) Electricity under strain, or Electrostatics; (2) Electricity in Locomotion, or current Electricity; (3) Electricity in Whirling Motion, or Magnetism; (4) Electricity in Vibration, or Radiation, commonly called Light. The propagation of electro-magnetic waves through space, the nature of what is called self-induction, the physical meaning involved in the terms magnetic permeability and specific inductive capacity, are all fully treated and explained by mechanical illustrations. The discoveries of Faraday, of Kerr, and of Hall; the velocity of electro-magnetic disturbances; the new discoveries relating to the transmission of electrical energy through the insulating medium instead of through a conductor, and other similar views associated to some extent with the names of Poynting and Heaviside; the magnetic observations of Ewing, the most recent discoveries of Hertz; are all treated and explained in a more or less full and popular manner. The treatment is adapted to that large class of persons who, having some acquaintance with the ordinary facts and phenomena of electrical science as detailed in the usual text-books, find some difficulty in perceiving the theoretical bearing of the whole, or in reading the works of the masters of science. The book begins by assuming an elementary knowledge of facts, gradually develops the "incompressible-fluid" idea of electricity, and thence leads on slowly to some of the most recent speculations and opinions concerning the structure of Ether, the nature of Light, the conceptions of Electricity, of Elasticity, and of Matter, and the relationship existing between them. It thus aims at placing its readers on a higher platform whence they can follow the still further progress which in our own day is being so rapidly accomplished in these difficult branches of Natural Science.

PREFACE.

THE doctrine expounded in this book is the etherial theory of electricity. Crudely one may say that as heat is a form of energy, or a mode of motion, so electricity is a form of ether, or a mode of etherial manifestation.

This doctrine is led up to by gradual stages, and the explanations in Part I. do not aim at the same fulness of detail as those in Part III. or IV. Since the book is intended to be useful to the higher class of students it seemed very permissible to adopt a method which I always use in teaching,—viz. to begin by giving *some* ideas at first, and to gradually polish them up later, rather than by attempting a too highly finished statement *ab initio* to overburden and depress, and possibly to confuse, a student. Because of this progressive arrangement I may be permitted to urge students to read the book through before proceeding to dip into it by help of the index and before taking notice of references forward, which subsequently it is hoped will prove useful.

Persons who are occupied with other branches of science or philosophy or with literature, and who have, therefore, not kept quite abreast of physical science, may possibly be surprised to see the intimate way in which the ether is now spoken of by physicists, and the assuredness with which it is experimented on. They may be inclined to imagine that it is still a hypothetical medium whose existence is a matter of opinion. Such is not the case. The existence of an ether can legitimately be denied in the same terms as the existence of matter can be denied, but only so. The evidence of its existence can be doubted or explained away in the one case as in the other, but the evidence for ether is as strong and direct as the evidence for air. The eye may indeed be called an etherial sense-organ, in the same sense as the ear can be called an aërial one, and somewhat in the same sense as the hand and muscles may be called a sense-organ for the appreciation of ordinary matter.

Some of the details of my explanations may be wrong (though I hope not), and all must be capable of ultimate improvement, but as to the main doctrine concerning the nature of electricity, though I call it a "view," it is to me no view but a conviction. Few things in physical science appear to me more certain than that what has so long been called electricity is a form, or rather a mode of manifestation, of the ether. Such words as "electrification," "electric," may remain; "electricity" may gradually have to go. It can be noticed that whereas in the earlier part of the book the word electricity occurs frequently and the word ether seldom, in the later portion this order of frequency is inverted.

A rough and crude statement adapted for popular use is that electricity and ether are identical; but that is not all that has to be said, for there are two opposite kinds of electricity, and there are not two ethers. But there may be two aspects of one ether, just as there are two sides to a sheet of paper, or two aspects of a transparent clock face; and similarly may positive and negative electricity be two aspects, or, as I have sometimes called them by chemical analogy, "components," of the ether. Anything which can be sheared (and ether is sheared by every electromotive force applied to it) must consist of two parts sufficiently different to travel or to be displaced in opposite directions.

If this statement is vague, it is because our present

knowledge of the structure of the ether is vague ; not because the relationship of electricity to ether is uncertain, or will be anything but definite so soon as we know the constitution of the ether more precisely. Vague at present our knowledge of the ether is, but not so vague as these lines may suggest.

That which has now to be investigated is not the nature of electricity, but the nature of the ether. Explanation always progresses by stages ; no explanation is ultimate ; every explanation is a step up, a removal of a thing from a lower to a higher category. Thus comets at one time might have been anything ; they have been shown to be a form (or swarm) of meteorites. Meteorites, again, have been shown to be lumps of common matter—usually iron or rock. There remains the question, What is iron or rock, or any form of matter ? Heat was once thought to be a form of matter ; it is now known to be a form of energy. There remains the question, What is energy ? Electricity has been thought to be a form of energy ; it has been shown to be a form of ether. There remains the question, What is ether ?

And a question it is indeed : *the* question of the physical world at the present time. But it is not unanswerable : it is, in my belief, not far from being answered. And it is probably a simpler question than the supplementary and next subsequent question, What is matter ? It is simpler, partly because ether is one, while matter is apparently many ; partly because the presence of matter so modifies the ether that no complete theory of the properties of matter can possibly be given without a preliminary and fairly complete knowledge of the properties and constitution of undisturbed ether in free space. When this has been attained, the resultant and combined effect we call matter may begin to be understood.

If a continuous incompressible perfect fluid filling all space can be imagined in such a state of motion that it will do all that ether is known to do ; if, simply by reason of its state of motion, it can be proved capable of conveying light and of manifesting all electric and magnetic phenomena which do not depend on the presence of matter ; and if the state of motion so imagined can be proved stable and such as can readily exist, the theory of free ether is complete.

The latest contribution towards such a theory appears while I write in a letter to *Nature* by G. F. Fitzgerald (May 9th, 1889). The fluid structure there imagined, a liquid in turbulent or vortex motion consisting of interlaced vortex filaments like a sponge, is proved to be capable of doing all that is required of free ether ; and the motion is believed with great probability to be a stable and possible state. The Fitzgerald ether may

consist, for instance, of an assemblage of columnar vortices
threading each other in three cardinal directions in square (or
cubical) order ; adjacent vortices rotating opposite ways like
the cells in my sectional diagrams, Figs. 37 and 46, in which
the clockwise whirls are positive electricity and the counter-
clockwise are negative electricity. Until we come into the
neighbourhood of matter no further distinction but exact
opposition of properties is existent. A somewhat similar idea
concerning the ether has been worked at by Mr. Hicks, see
§ 156 ; and Sir William Thomson proved in a famous paper at
the British Association in 1887, that a laminar arrangement of
vortices could transmit transverse vibrations (*i.e.* light), though
with some absorption and therefore partial opacity (*Phil. Mag.*
October 1887). Fitzgerald has now gone a step further and
devised a fibrous ether which is not only optically, but also
electrically, sufficient. If no flaw appears, if it stand the test
of criticism and further development, the theory of free ether is
far more than begun.

The theory of bound ether and of matter must next
follow, and thereby, in addition to all optical and
electrical phenomena, gravitation and cohesion must
be explained too. Then must be attacked the specific
differences between various kinds of matter, and the
nature of what we call their " combinations." When
this is accomplished the complex facts of chemistry
will have been brought under a comprehensive law.

The next fifty years may witness these tremendous
victories in great part won.

UNIVERSITY COLLEGE, LIVERPOOL.
May 13th, 1889.

CONTENTS.

PART I.

INTRODUCTION AND ELECTROSTATICS.

PART II.

CONDUCTION.

CHAPTER IV.

CHAPTER V.

CHAPTER VI.

PART III.

MAGNETISM.

CHAPTER VII.

CHAPTER VIII.

CHAPTER IX.

PAGE

CHAPTER X.

CHAPTER XI.

PART IV.

RADIATION.

CHAPTER XII.

CHAPTER XIII.

CHAPTER XIV.

CHAPTER XV.

APPENDED LECTURES

PART I.

INTRODUCTION AND ELECTROSTATICS.

CHAPTER I.

1. IT is often said that we do not know what electricity is, and there is a considerable amount of truth in the statement. It is not so true, however, as it was some twenty years ago. Some things are beginning to be known about it ; and though modern views are tentative, and may well require modification, nevertheless some progress has been made. I shall endeavour in this essay to set forth as best I may the position of thinkers on electrical subjects at the present time.

I begin by saying that the whole subject of electricity is divisible for purposes of classification into four great branches.

(1) Electricity at rest, or static electricity : wherein are studied all the phenomena belonging to stresses and strains in insulating or dielectric media brought

B 2

about by the neighbourhood of electric charges or electrified bodies at rest immersed therein ; together with the modes of exciting such electric charges and the laws of their interactions.

(2) Electricity in locomotion, or current electricity : wherein are discussed all the phenomena set up in metallic conductors, in chemical compounds, and in dielectric media, by the passage of electricity through them ; together with the modes of setting electricity in continuous motion and the laws of its flow.

(3) Electricity in rotation, or magnetism : wherein are discussed the phenomena belonging to electricity in whirling or vortex motion, the modes of exciting such whirls, the stresses and strains produced by them, and the laws of their interaction.

(4) Electricity in vibration, or radiation : wherein are discussed the propagation of periodic or undulatory disturbances through various kinds of media, the laws regulating wave velocity, wave-length, reflection, interference, dispersion, polarization, and a multitude of phenomena studied for a long time under the heading "Light." Although this is the most abstruse and difficult portion of electrical science, a certain fraction of it has been known to us longer than any other branch, and has been studied under special advantages, because of our happening to possess a special sense-organ for its appreciation.

Now in order to be able to get through a survey of these four great and comprehensive groups in moderate compass, it will be necessary for me to assume acquaintance with all the elementary facts and proceed at once to their elucidation.

2. The great names in connection with our progress in knowledge as to the real nature of electricity, irrespective of a mere study and extension of its known facts, are

FRANKLIN, CAVENDISH, FARADAY, MAXWELL.

To these, indeed, you may feel impelled to add the tremendous name of THOMSON ; but one has some delicacy in attempting to estimate the work of living philosophers, and as Maxwell has been very explicit in acknowledging his indebtedness to his illustrious contemporary, whose work will in the course of nature have to be criticized and appraised by far abler hands than mine and by the philosophers of generations yet unborn, we may well afford to abstain from minute considerations and accept for the present the name of Maxwell as representative of the great English school of mathematical physicists, under whose influence, Cambridge, in the pride of having reared them, is awaking to new and energetic scientific life, and whose splendid achievements will shine out in the future as the glory of this century.

The views concerning electrification which I shall try to explain are in some sense a development of those originally propounded by that most remarkable man, Benjamin Franklin. The accurate and acute experimenting of Cavendish laid the foundation for the modern theory of electricity ; but, as he worked for himself rather than for the race, and as moreover he was in this matter far in advance of his time, Faraday had to go over the same ground again, with extensions and additions peculiar to himself and corresponding to the greater field of information at his disposal three-quarters of a century later. Both these men, and especially Faraday, so lived among phenomena that they yielded up their hidden secrets to them in a way unintelligible to ordinary workers ; but while they themselves arrived at truth by processes that savour of intuition, they were unable always to express themselves intelligibly to their contemporaries and to make the inner meaning of their facts and speculations understood. Then comes Maxwell, with his keen penetration and great grasp of thought combined with mathematical subtlety and power of expression ; he assimilates the facts, sympathizes with the philosophic but untutored modes of expression invented by Faraday, links the theorems of Green and Stokes and Thomson to the facts of

Faraday, and from the union there arises the young modern science of electricity, whose infancy at the present time is so vigorous and so promising that we are all looking forward to the near future in eager hope and expectation of some greater and still more magnificent generalization.

3. You know well that there have been fluid or material theories of electricity for the past century ; you know, moreover, that there has been a reaction against them. There was even a tendency a few years back to deny the material nature of electricity and assert its position as a form of energy. This was doubtless due to an analogical and natural, though unjustifiable, feeling that just as sound and heat and light had shown themselves to be forms of energy so in due time would electricity also. If such were the expectation, it has not been justified by the event. Electricity may possibly be a form of matter—it is not a form of energy. It is quite true that electricity *under pressure* or *in motion* represents energy, but the same thing is true of water or air, and we do not therefore deny them to be forms of matter. Understand the sense in which I use the word electricity. *Electrification* is a result of work done, and is most certainly a form of energy ; it can be created and destroyed by an act of work. But electricity—none is ever created or destroyed, it is simply moved and strained like matter. No one ever exhibited a trace

of positive electricity without there being somewhere in its immediate neighbourhood an equal quantity of negative.

The simplest proof of this statement consists in making experiments inside a closed conducting insulated room or shell ; it may be the size of a living room, or the size of a beer-tankard, whichever is most convenient. All known electrical experiments being performed inside such a room, bodies electrified strongly, moved about, sparks taken, &c., &c., a sensitive electroscope connected to the room shall not show the slightest permanent effect. In other words the room will not become in the slightest degree charged. I say no *permanent* effect, because it is just possible that small transitory effects may occur during the rapid rearrangement of internal charges. I do not feel sure whether such transitory effects are or are not really possible, but whether they are or not, they have nothing to do with our present argument. All the electrification that have been going on will not have resulted in the creation of the minutest quantity of electricity ; the only way to charge the room is to pass a charge in from some other body outside.

This is the first great law, expressible in a variety of ways : as, for instance, by saying that total algebraic production of electricity is always zero ; that you cannot produce positive electrification without an equal quantity of negative also ; that what one body gains of electricity some other body must lose.

Now, whenever we perceive that a thing is produced in precisely equal and opposite amounts, so that what one body gains another loses, it is convenient and most simple to consider the thing not as generated in the one body and destroyed in the other, but as simply *transferred. Electricity in this respect behaves just like a substance.* This is what Franklin perceived.

4. The second great law is that electricity always, under all circumstances, flows in a closed circuit, the same quantity crossing every section of that circuit, so that it is not possible to exhaust it from one region of space and condense it in another.

Another way of expressing this fact is to say that no charge resides in the interior of a hollow conductor, but that every trace of charge is on the outer surface and penetrates to no appreciable depth.

Another is to say that total induced charge is always equal and opposite to inducing charge.

This second law can also be illustrated by the insulated room or conducting cavity already mentioned. Having found that internal electrification produces no effect on an outside electroscope connected to the walls, proceed to pass a charge in to the cavity through a temporarily opened window or lid. Instantly the chamber has become charged by a definite amount, viz. by the precise amount which has then been intro-

duced into its cavity.   The charge need not be in any
way communicated to the chamber, all that is necessary
is that it shall be wholly inside.   Moving the charge
about, or letting it spark to the walls of the chamber,
makes not the slightest difference to the electroscope outside.   It may be watched through a micro-
scope at the instant the spark occurs and it will not
show the slightest twinkle.   Both these experiments
of the hollow chamber were made by Faraday, and the
latter is well known and often quoted as his " ice-pail "
experiment, because he happened to use an ice-pail
sometimes as his insulated chamber.

Another mode of illustrating the same series of facts
is afforded by an insulated parrot cage with an electroscope inside it connected by a wire to the bars of the
cage.   The cage may now be charged strongly, its
potential may be changed from a million volts positive
to a million volts negative, sparks of any length may
be taken from it ; but provided the meshes are close
enough for it to be regarded as a really closed vessel
the electroscope inside is wholly unaffected.   In
making this experiment the electroscope must be
connected with the bars of the cage, for if it be detached it can be affected by charged air blown through
the meshes.   Directly a charge gets *inside* the cage it
can affect the electroscope easily enough unless there
is a wall-connexion, and then it is powerless.

The open-work of a cage is objectionable on this

account, that it allows the permeation of electrified air, just as it might allow an electrified pith ball to be thrown in. Making the meshes close is no guarantee against this source of disturbance : I have blown electrified air from a point through very fine earth-connected copper gauze and affected strongly an electroscope on the other side.

No doubt a solid metal walled room is secure against this cause of disturbance but then it is difficult to see the electroscope. Faraday however constructed a room for the purpose big enough to get into himself, and thus performed the experiment quite satisfactorily.

But perhaps the most rigorous mode of examining the precise truth of the property of electricity which lies at the bottom of all these experiments is that adopted in the famous Cavendish experiment : sometimes referred to in French books by the name of Biot. This consists in charging strongly an insulated sphere provided with a couple of hemispherical caps which can afterwards be taken off mechanically, and connecting a delicate electroscope immediately after to the disclosed ball. Not the faintest trace of charge is found upon it. This experiment has been repeated by Clerk Maxwell in the Cavendish Laboratory, Cambridge, using a Thomson Quadrant Electrometer and all modern appliances, with absolutely negative result.[1]

---

[1] See Maxwell's " Electrical Researches of Cavendish," pp. 104 and 417. An interesting little experiment with soap bubbles, made by Mr.

This series of experiments are most vital, and give us most fundamental information regarding electricity : let us consider how we can best express their teaching.

5. When we thus find that it is impossible to charge a body absolutely with electricity, that though you can move it from place to place it always and instantly refills the body from which you take it, so that no portion of space can be more or less filled with it than it already is, that it is impossible by any rise of potential to squeeze a trace of electricity into the interior of a cavity, and that if a charge be introduced a precisely equal quantity at once passes through the walls to the outside ; it is natural to express the phenomenon by saying that electricity behaves itself like a perfectly *incompressible* substance or fluid, of which all space is completely full. That is to say, it behaves like a perfect and all permeating *liquid.* Understand I by no means assert that electricity *is* such a fluid or liquid ; I only assert the undoubted fact that it behaves like one, *i.e.* it obeys the same laws.

It may be advisable carefully to guard oneself against becoming too strongly imbued with the notion that because electricity obeys the laws of a liquid

Vernon Boys, illustrates the fact that the depth to which a charge penetrates is less than the diameter of a few molecules, for one soap bubble inside another is entirely screened from such electrostatic forces as can be applied.

therefore it is one. One must always be keenly on the lookout for any discrepancy between the behaviour of the two things, and a single contradictory discrepancy—not a mere difference but a real opposition of properties—will be sufficient to overthrow the fancy that they may perhaps be really identical. Till such a discrepancy turns up, however, we are justified in pursuing the analogy—more than justified, we are impelled. And if we resist the help of an analogy like this there are only two courses open to us : either we must become first-rate mathematicians, able to live wholly among symbols and dispensing with pictorial images and such adventitious aid ; or we must remain in hazy ignorance of the stages which have been reached, and of the present knowledge of electricity so far as it goes. I need hardly say that by " modern views " I do not mean *ultimate* views ; nor do I mean that I can give an account of all the speculations and ideas floating in the minds of some two or three of our most advanced thinkers. All I attempt is to give an account of the stage which has certainly been attained, to indicate the directions in which immediate progress is probable, and to ask you to take for granted that the next quarter of a century will see as great advances made upon these views as they are superior to the doctrines inculcated by the ordinary run of text-books.

6. Imagine now that we live immersed in an

infinite ocean of incompressible and inexpansible all-permeating perfect liquid, like fish live in the sea : how can we become cognizant of its existence ? Not by its weight, for we can remove it from no portion of space in order to try whether its has weight.

We can weigh air, truly, but that is simply because we can compress it and rarefy it. An exhausting or condensing pump of some kind was needed before even air could be weighed or its pressure estimated.

But if air had been incompressible and inexpansible, if it had been a vacuum-less perfect liquid, pumps would have been useless for the purpose, and we should necessarily be completely ignorant of the weight and pressure of the atmosphere.

How then should we become cognizant of its existence ?   In four ways :—

(1) By being able to pump it out of one elastic bag into another [not out of one bucket into another : if you lived at the bottom of the sea you would never think about filling or emptying buckets—the idea would be absurd ; but you could fill or empty elastic bags], and by noticing the strain phenomena exhibited by the bags and their tendency to burst when over full. [Water (or air) was here pumped out of one elastic bag into another, and the analogy with an electrical machine charging two conductors oppositely, *i.e.* pumping electricity from one into the other, was pointed out.]

(2) By winds or currents ; by watching the effect of moving masses of the fluid as it flows along pipes or through spongy bodies, and by the effects of its inertia and momentum. [A hanging vane arranged in a tube so as to be deflected by a stream of water was here likened roughly to a galvanometer ; also the effect of suddenly stopping a stream of water, as in a water ram, was mentioned as analogous to self-induction.]

(3) By making vortices and whirls in the fluid, and by observing the mutual action of these vortices, their attractions and repulsions. [Whirlwinds, sand-storms, waterspouts, cyclones, whirlpools.]

(4) By setting up undulations in the medium : i.e. by the phenomena which in ordinary media excite in us through our ears the sensation called " *sound*."

In all these ways we have become acquainted with electricity, and in no others that I am aware of. They correspond to the four great divisions of the subject which I made above.

7. But there are differences, very important differences, between the behaviour of a material liquid ocean such as we have contemplated and the behaviour of electricity. First it is doubtful whether electricity by itself and disconnected from matter has any inertia. It is by no means certain that it has not : the experiments made by Maxwell with a negative result (§ 39) need only prove either that its

speed of flow is very small, or that an electric current consists of equal opposite streams of equal momentum. The laws of electric flow in conductors are such as indicate no inertia, (§ 48) and this fact would be conclusive were it not that a recent brilliant paper by Prof. Poynting explains the reason of it completely otherwise, and leaves the question of inertia quite open ; on the other hand, the facts of magnetism seem definitely to require inertia, or something corresponding to it. Leaving this for the present as an open question, there can be no doubt but that when in connection with insulating or dielectric matter *the combination* most certainly possesses inertia (§§ 38 and 39).

8. A more serious and certain difference between the behaviour of electricity and that of an incompressible fluid comes out in the fourth category—that concerned with wave-motion. In an incompressible fluid the velocity and length of waves would both be infinite, and none of the phenomena connected with the gradual propagation of waves through it could exist. Such a medium therefore would be incapable of sound-vibrations in any ordinary sense. On the other hand, it is quite certain that the disturbances concerned in light-radiation take place at right angles to the direction of propagation—they are transverse disturbances—and such disturbances as these no body with the entire properties of a fluid can possibly transmit. Remember, however, that the medium

which transmits light is the ether and not simply electricity. We have nowhere asserted that electricity and the ether are identical. If they are, we are bound to admit that ether, though fluid in the sense of enabling masses to move freely through it, has a certain amount of rigidity for enormously rapid and minute oscillatory disturbances. If they are not identical we can more vaguely say that ether contains electricity as a jelly contains water, but that the rigidity concerned in the transverse vibrations belongs not to the water in the jelly but to the mode in which it is entangled in its meshes. However all this is a great and difficult question into which we shall be able to enter with more satisfaction twenty years hence (Chap. XII.).

Provisionally we will accept as a temporary working hypothesis the idea of the ether consisting of electricity in a state of entanglement similar to that of water in jelly ; and we are driven to this view by the exigencies of mode 1 (§§ 1 and 6), the electrostatic or strain method of examining the properties of electricity, because otherwise the properties of insulators are hard to conceive. If it turn out that space is a conductor, which seems to me highly improbable, not to say absurdly impossible, then we must fall back upon the other view that it is rigid only for infinitesimal vibrations, and fluid for steady forces.

C

CHAPTER II.

9. RETURN now to the consideration of electrostatics. We are to regard ourselves as living immersed
in an infinite all-permeating ocean of perfect incompressible fluid (or liquid), as fish live in the sea ; but this
is not all, for if that were our actual state we should have
no more notion of the existence of the liquid than deep-
sea fish have of the medium they swim in. If matter
were all perfectly conducting, it would be our state :
in a perfectly free ocean there is no insulation—no
obstruction to flow of liquid : it is the fact of insulation that renders electrostatics possible. We could
obstruct the flow and store up definite quantities of a
fluid in which we were totally submerged by the use
of closed vessels of course. But how could we pump
liquid from one into another so as to charge one
positively and another negatively ? Only by having
the walls elastic : by the use of elastic bags, and
elastic partitions across pipes. And so we can represent
a continuous insulating medium (like the atmosphere

or space) by the analogy of a jelly, through which liquid can only flow by reason of cracks and channels and cavities.

Modify the idea of an infinite ocean of liquid into that of an infinite jelly or elastic substance in which the liquid is entangled, and through which it cannot penetrate without violence and disruption ; and you have here a model of the general insulating atmosphere.   Our ocean of fluid is not free and mobile like water, it is stiff and entangled like jelly.

Nevertheless bodies can move through it freely. Yes, *bodies* can, it is the *liquid* itself only which is entangled.   How we are to picture freely and naturally the motion of ordinary matter through the insulating medium of space it is not easy to say.   It is a difficulty not fatal, but sensible, and due to an imperfection in our analogy.

Insulators being like elastic partitions or impervious but yielding masses, conductors are like cavities, porous or spongy bodies perfectly pervious though with more or less frictional resistance to the flow of liquids through them.   Thus, whereas bodies easily penetrable by matter are impervious to electricity, bodies like metals which resist entirely the passage of matter, are quite permeable to electricity.   It is this inversion of ordinary ideas of penetrability that constitutes a small difficulty at the beginning of the subject.

However, supposing it overcome, let us think of

C 2

these insulated spheres and cylinders on the table connected by copper wire as so many cavities and tubes in an otherwise continuous elastic impervious medium which surrounds them and us, and extends throughout space wherever conductors are not. All, however, cavities as well as the rest of the medium, are completely full of the universal fluid. The fluid which is entangled in insulators is free to move in conductors ; whence it follows that its pressure or potential is the same in every part of a conductor in which it is not flowing along. For if there were any excess of pressure at any point, a flow would immediately occur until it was equalized. In an insulator this is by no means the case. Differences of pressure are exceedingly common in insulators, and are naturally accompanied by a strain of the medium.

It is instructive now to think over a number of ordinary electrical experiments from this point of view : to think of the fluid as flowing freely through conductors and settling down to a state of equilibrium or uniform pressure in them, but straining insulators as high pressure water might strain elastic walls or boundaries, straining them even to bursting if the partitions be made too thin.

10. There have been, as you know, two ancient fluid theories of electricity—the one-fluid theory of Franklin, and the two-fluid theory of Symmer and others. A great deal is to be said for both of

them within a certain range. There are certainly points, many points, on which they are hopelessly wrong and misleading, *but it is their foundation upon ideas of action at a distance that condemns them, it is not the fluidity.* They concentrate attention upon the conductors ; whereas Faraday taught us to concentrate attention on the insulating medium surrounding the conductors—the *"dielectric"* as he termed it. This is the seat of all the phenomena : conductors are mere breaks in it—interrupters of its continuity.

To Faraday the space round conductors was full of what he called lines of force ; and it is his main achievement in electrostatics to have diverted our attention from the obvious and apparent, to the intrinsic and essential, phenomena. Let us try and seize his point of view before going further. It is certainly true as far as it goes, and is devoid of hypothesis.

Take the old fundamental electric experiment of rubbing two bodies together, separating them, and exhibiting the attraction and repulsion of a pith ball, say ; and how should we now describe it ? Something this way :—

Take two insulated disks of different material, one metal, say, and one silk, touch them together ; the contact effects a transfer of electricity from the metal to the silk ; rub slightly to assist the transfer, since silk is a non-conductor, then separate. As you

separate the disks the medium between them is thrown
into a state of strain, the direction of which is mapped
out by drawing a set of lines, called lines of force,
from one disk to the other, coincident with the direction of strain at every point.   As Faraday remarked,
the strain is as if these lines were stretched elastic
threads endowed with the property of repelling each
other as well as of shortening themselves ; in other
words, there is a tension along the lines of force and
a pressure at right angles to them.   When the disks
are near, and the lines short, they are nearly straight

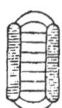

FIG. 1.—Rough diagram of the state of the medium near two oppositely charged
disks when close together.

(Fig. 1), but as the distance increases they become
curved, bulging away from the common axis of the
two disks and some even curling round to the back of
the disk (Fig. 2), until when the disks are infinitely
distant as many lines spring from the back of each as
from its face ; and we have a charged body to all
intents existing in space by itself.

The state of tension existing in the medium between
the disks results in a tendency to bring them together
again, just as if they were connected by so many
elastic threads of no length when unstretched.   The

ends of the lines are the so-called electrifications or
charges, and the lines perpetually try to shorten and
shut up, so that their ends may coincide and the strain
be relieved.   If one of the disks touch another conducting body, some of its lines instantly leave it and
go to the body ; in other words, the charge is capable
of transference, and the new body is urged towards

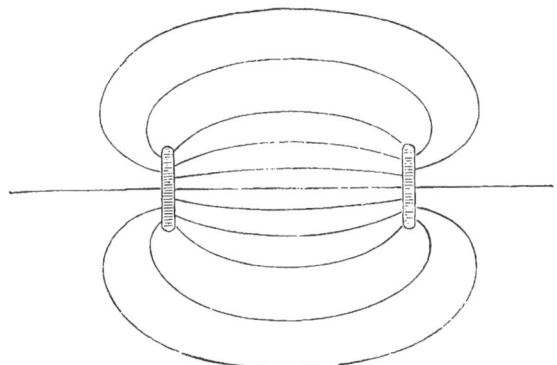

FIG. 2.—Rough diagram of the state of the medium near two oppositely charged
disks when separated.

the other disk, just as the disk was from which it
received the lines.   If this new body *completely surrounds* the disk, it receives the whole of its lines, and
the disk can be withdrawn perfectly free and inert.
[Faraday's " ice pail " experiment (§ 4).]

11. Now take the two charged disks, facing one
another, and let, say, a suspended gilt pith ball hang
between them.   Being a conductor there is no strain

inside it, and so it acts partially as a bridge, and several of the lines pass through it—or, rather, they end at one side of it and begin at the other: thus it has opposite charges on its two faces—it is under induction (Fig. 3). Let it now be moved so as to touch one of the disks, the lines between it and the disk on that side have shut up, and it remains with

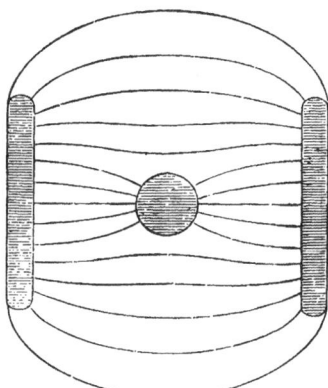

FIG. 3.—Rough diagram of the medium between two disks disturbed by the presence of an uncharged metal sphere. The two halves of the sphere are oppositely charged "by induction."

those only which go to the other disk.   In other words, it has received, unbalanced, some of the lines which previously belonged to the touched disk. These will pull it over to the far disk and there shut themselves up.   From that disk it receives more ; and it travels with their ends back to the first disk, and so on (Fig. 4), perpetually receiving lines and shutting them up, until they are all gone and the disks are discharged.

This mode of stating the facts involves no hypothesis whatever—it is the simple truth. But the

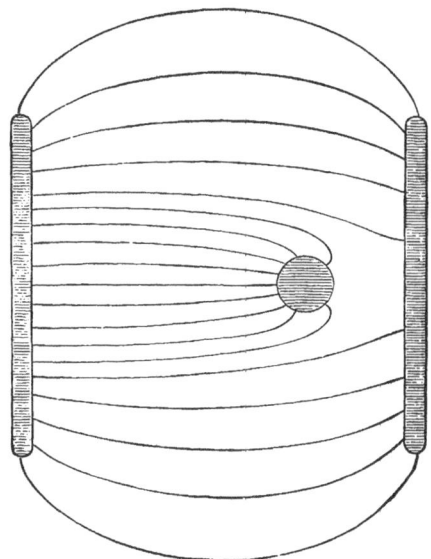

FIG. 4.—Rough diagram of the medium near two oppositely charged disks between which a metal carrier ball is oscillating, having just touched the right-hand disk. (Discharge by "alternate contact.")

" lines of force" have no more and no less existence than have "rays of light." Both are convenient modes of expression.

12. But so long as we adhere to this mode of

expression we cannot form a complete mental picture of the actually occurring operations. In optics it is usual to abandon rays at a certain stage and attend to the waves, which we know are of the essence of the phenomenon, though we do not yet know very much about their true nature.

Similarly in electricity, at a certain point we are led to abandon lines of force and potential theories, and to try to conceive the actual stuff undergoing its strains and motions. It is then we get urged towards ideas similar to those which are useful in treating of the behaviour of an incompressible fluid.

In an utterly modified sense, we have still a fluid theory of electricity, and a portion of the ideas of the old theories belong to it also.

Thus Franklin's view that positive charge was excess, and negative charge was a deficit, in a certain standard quantity of the fluid which all bodies naturally possessed in their neutral state, remains practically true. His view that the fluid was never manufactured, but was taken from one body to give to another, so that one gained what the other lost— no more and no less—remains practically true. Part also—a less part—of the two-fluid theory likewise remains true, in my present opinion (§ 90) ; but this is not a branch of the subject on which I shall enter in the present part. It will suffice for the present to fix our attention on one fluid only.

You are to think of an electric machine as a pump which, being attached to two bodies respectively, drives some electricity from the one into the other, conferring upon one a positive and upon the other a precisely equal negative charge. One of the two bodies may be the earth, in which case the charge makes little or no difference to it.

13. But, as has been objected before, if electricity is like an incompressible and inextensible fluid, how is it possible to withdraw any of it from one body and give it to another? With rigid bodies it is not possible, but with elastic bodies it is easy.

The act of charging this sphere is therefore analogous to pumping water into this elastic bag, or rather into a cavity in the midst of an elastic medium, whose thick walls, extending in all directions and needing a great pressure to strain them, better represent the true state of the case than does the thin boundary of a bag like this.

Draw a couple of such cavities and consider fluid pumped from one into the other, and you will see that the charge (i.e. the excess or defect of fluid) resides on the outside.

If the fluid is exactly incompressible, not the least extra quantity will be squeezed by the pressure into the space originally occupied by the cavity. This is the moral of the Cavendish experiment (§ 4): it proves that electricity is precisely incompressible.

You may also show that when both cavities are similarly charged the medium is so strained that they tend to be forced apart ; whereas when one is distended and the other contracted they tend to approach.

Further you may consider two cavities side by side, pump fluid into (or out of) one only, and watch the effect on the other. You will thus see the phenomena of induction, the near side of the second cavity becoming oppositely charged (*i.e.* the walls encroaching on the cavity), the far side similarly charged (the cavity encroaching on the walls), and the pressure on the fluid in the cavity being increased or diminished in correspondence with the rise or fall of pressure in the charged or inducing cavity. In other words, conductors rise in potential when brought near a positively charged body ; and their charge, if any, though not altered in quantity becomes redistributed.

The actual changes in volume necessary to the strain of these cavities are a defect in the analogy. To avoid this objection, one will have to accept a dual view of electricity—a sort of two-fluid theory, which many phenomena urge one to accept, but about which I will say nothing at present. It is sufficient at first to grasp the one-fluid ideas (§ 90).

14. *Return Circuit.*—Sometimes a difficulty is felt about electricity flowing in a closed circuit—as, for instance, in signalling to America and using the earth

as a return circuit : the question arises, How does the electricity find its way back ?

The difficulty is no more real than if a tube were laid to America with its two ends connected to the sea and already quite full. If now a little more sea-water were pumped in at one end, an equal quantity would leave the other end, and the disturbed level of the ocean would readjust itself. Not the same identical water would return, but an equal quantity would return. That is all one says of electricity. One cannot label and identify electricity.

To imitate the inductive retardation of cables, the tube should have slightly elastic walls ; to imitate the speed of signalling, the water must be supposed quite incompressible, not elastic as it really is, or each pulse would take three-quarters of an hour to go.

CHAPTER III.

CHARGE AND INDUCTION.

15. *Condensers.*—Returning to the subject of charging bodies electrically, how is one to consider the fact that bringing an earth-plate near a conductor increases its capacity so greatly, enabling the same pressure to force in a much larger quantity of fluid? how is one to think of a condenser, or Leyden jar?

In the easiest possible way, by observing that the bringing near an earth-connected conductor is really *thinning down the dielectric* on all sides of the body.

The thin-walled elastic medium of course takes less force to distend it a given amount than a thick mass of the same stuff took; in other words it has much more capacity. Remember that capacity of elastic cavities cannot satisfactorily be measured as the capacity of buckets is measured, by the maximum quantity they will hold when full, they are never

"full" till they burst ; and the amount required to
burst them measures rather their strength than their
capacity.   The only reasonable definition of capacity
in such cases is the ratio of any addition to their
contents to the extra pressure required to force it in :
and this is exactly the way electrical "capacity" is
defined.   A Leyden jar is like a cavity with quite
thin walls—in other words, it is like an elastic
bag.

But if you thin it too far, or strain it too much, the
elastic membrane may burst : exactly, and this is the
disruptive discharge of a jar, and is accompanied by
a spark.   Sometimes it is the solid dielectric which
breaks down permanently.   Ordinarily it is merely
the air ; and, since a fluid insulator constitutes a
self-mending partition, it is instantaneously as good
as new again.

There are many things of interest and importance
to study about a Leyden jar.   There is the fact that
if insulated, it will not charge : the potential of both
inner and outer coatings rises equally ; that, in order
to charge it, for every positive spark you give to the
interior an equal positive spark must be taken from
the exterior.   There is the charging and the discharging of it by alternate contacts, as by an oscillating
ball ; and there are the phenomena of the spark-
discharge itself.

But, as you know, *all* charging is really a case of a

Leyden jar. The outer coat must always be somewhere—the walls of the room, or the earth, or
something—you always have a layer of dielectric
between two charges—the so-called induced and
the inducing charge. You cannot charge one body
alone (§ 5).

16. To illustrate the phenomena of charge, I will
now call your attention to these diagrams—which
less completely but more simply than hydraulic
illustrations, serve to make the nature of the
phenomena manifest.

First you have an inextensible endless cord circulating over pulleys ; this is to represent electricity
flowing in a closed circuit. Electromotive forces are
forces capable of moving the cord, and you may
consider them applied either by a winch, or by a
weight on the hook W. A battery cell corresponds to
a small weight ; an electric machine to a slow but
powerful winch. Clamping the cord with the screw S
corresponds to making the resistance of the circuit
infinite. Instead of the cord, clamp, and driving
pulley, one might consider an endless pipe full of
liquid with a stop-cock and a pump on it, but for
many purposes the cord is sufficient and more simple.
In Fig. 5, the only resistance to the motion is friction,
and there is no tendency to spring back. Fixed
beads are shown on the cord to typify atoms of
matter, and they may be more or less rough to

represent different specific resistances.   If the cord be moved, heat is the only result.

Now pass to Fig. 6.   Here the cord is the same as before, but the beads are firmly attached to it, so that if it moves they must move with it.   They represent, therefore, the particles of an insulating substance.

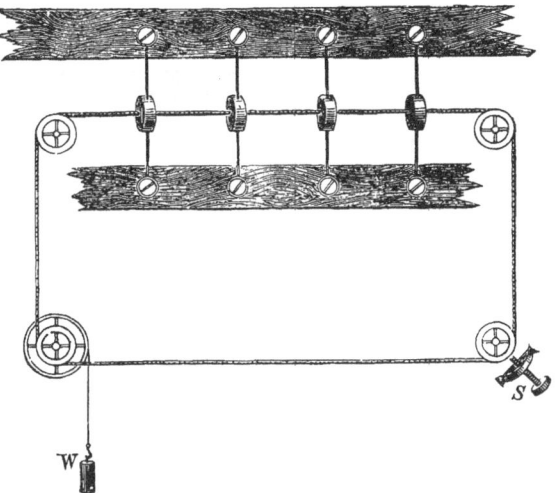

FIG 5.—Mechanical analogy of a *metallic* circuit.

Nevertheless, their supports are not rigid—they do not prevent the cord moving at all ; they allow what is called electric "*displacement*," not conduction ; they can be displaced a little from their natural position, but they spring back again when the disturbing E.M.F. is removed.   The beads in this

D

figure are supposed to be supported by elastic
threads : if the cord were replaced by a closed pipe
full of water the beads and threads would be replaced
by elastic partitions. The specific inductive capacity
of the dielectric is represented by the stretchability, or
inverse elasticity, of the elastic threads. The stiffer
the threads are to pull out, the less is the inductive

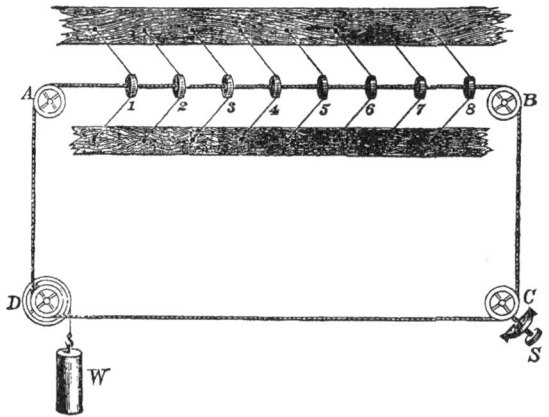

FIG. 6.—Mechanical analogy of a circuit partly *dielectric :* for instance, of a
charged condenser. A is its positive coat, B its negative.

capacity of the medium ; because evidently a greater
E.M.F. is needed to cause a given displacement.

Apply a given E.M.F. to this cord, as for instance
the weight w, and a definite displacement is produced.
One side, A, gets more cord than usual—it is
positively charged ; the other side, B, gets less—it is

negatively charged.   If the applied E.M.F exceeds a certain limit the strain is too great ; the elastics break, and you have disruptive discharge with a spark.   But even when the strain is only moderate some of the supports may yield viscously, or be imperfectly elastic and permit a gradual extra displacement of the cord, known to telegraphists as " soaking in."

When discharge is now allowed, it will not at once be complete ; a large portion of the displacement will be at once recovered, but the rest will gradually " soak out " and cause residual discharges.

If the dielectric is at all stratified in structure, so that some of the beads allow cord to slip through them—or yield more than others—then this residual charge effect will become very prominent.

17. These are matters which it is easy to thoroughly understand, and Fig. 7 illustrates different stages sufficiently.   In Diagram I. are represented 8 strata, each displaced from its normal position by an amount 3.   The restoring force being proportional to the displacement, the total restoring force can be called 24. The diagram represents, therefore, a Leyden jar or other dielectric strained by an applied E.M.F. of 24 units.   If every stratum insulates perfectly—that is, if every bead is quite firmly attached to the cord—nothing further happens so long as this force is kept applied. This state of things may be maintained in two ways, either by keeping on the weight w—that is, by keeping

the condenser permanently connected to the battery; or by clamping the cord and thus making the resistance infinite—that is, by insulating the terminals of the condenser.

But now suppose that some of the strata are not perfectly insulating ; let some of the beads slowly slide back along the cord towards their zero position. Then we shall witness different phenomena according to whether the weight has been left on, or the cord has been clamped.

Take first the case of leaving the weight on—that is, keeping the battery connected. If every bead slide equally, all we get is a continuance of the state represented in Diagram I, combined with a slow oozing forward of the cord—that is, a slow and steady leak through the condenser. But suppose every bead does not slide equally, suppose some do not slide at all ; then the slipping of some throws extra strain on the others, and the cord moves forward, but more and more slowly until the insulating strata ultimately have to bear all the strain, and the cord asymptotically comes to rest.

This process is observed in Atlantic cables and Leyden jars ; one is liable to it with all condensers except air condensers, and it is called " soaking in " ; it is accompanied by the development of internal charges, because plainly the original normal length of cord between the beads is no longer maintained ; some layers have acquired an extra length—that is, are positively charged—others are negatively charged.

The strain is distributed very unequally, but its total amount, in this case, continues constant.

Remove the electromotive force now : we get first a quick discharge, and then a slow leaking out or reverse motion of the cord as it is propelled by the still displaced insulating strata through the now oppositely displaced badly conducting ones ; the time of " soaking out " being comparable with that permitted to the soaking in.

It is important to see how these phenomena are entirely reconcilable with the incompressible character of electricity—that is, the unyielding character of the cord. The ordinary notion of the positive and negative charges of a Leyden jar crawling into the glass to meet each other, and then crawling out again, is quite erroneous ; and the actual process which simulates the effect of this impossible process is perfectly clear.

So much for the case when the battery is left connected ; now attend to the case when the terminals are insulated.

Having got the dielectric into the state represented by I., screw down the clamp and wait. If some of the beads slide, while some do not, we shall shortly arrive at the state represented by II. Beads Nos. 3, 5, and 6, have slid partially back, and the total strain

on the cord has been reduced to 17. The jar will
appear to have partially discharged itself by internal

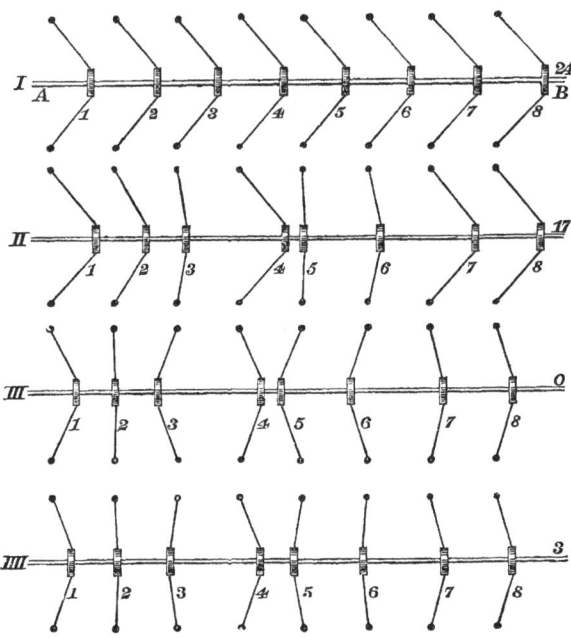

FIG. 7.—Stages in the discharge of a stratified condenser, with some of its layers of
        imperfect insulating power; showing one way in which the phenomena of
        "residual charge," "internal charge," and "soaking out" are produced.
   I. represents a recent charge, of E.M.F. 24.
   II. represents the same after lapse of time, reduced to 17 by partial internal leakage,
        and shows internal charge. The circuit itself is supposed to have been perfectly
        insulating all the time; the charge on the plates therefore remains constant.
   III. shows the first discharge.
   IIII shows the state attained after again waiting, viz. a residual charge with an
        E.M.F. 3 in the old direction.

leakage, and yet not the slightest motion of the cord
has been permitted. Internal charges have appeared

positive between Nos. 3 and 4 and between 5 and 6, negative between 2 and 3 and between 4 and 5. The charges on the coatings at A and B have remained constant the jar has apparently increased in capacity, because the same charge is maintained by a less electromotive force. All these effects may present themselves at first sight as irreconcilable with the behaviour of an incompressible fluid ; but the diagram clearly says otherwise.

Now unclamp the cord momentarily, *i.e.* discharge the condenser and insulate again. At the instant of discharge a rush of electricity takes place, and the force falls to zero. The state of the discharged jar at the first instant after discharge is represented in Diagram III. The surface charges have not wholly disappeared ; the internal charges have been unaffected ; the displacement of none of the strata is zero. The insulating ones remain with some of their original displacement, the leaking ones have been forced into a position of inverse displacement, so as just to reduce the force on the cord to zero. The most slippery one, No. 5, has now been most displaced in the inverse direction. But not long do they thus remain. They at once begin to slowly ooze back, and before long they will have got into the state represented by IV., where the now almost unbalanced stress of the insulating strata exerts on the cord a force 3 in the original direction. This is

known as the residual charge, and on unclamping
the cord, the first residual discharge can be obtained.
Not even yet, however, is the jar wholly discharged.
Waiting again, another but feebler residual charge
makes its appearance, and so on, almost without
limit, until the sum of all the residual discharges plus
the original discharge make up exactly the charge
originally imparted to the jar—make it up exactly
if any one of the strata has declined to leak. If all
leaked more or less, then of course there will be
some deficiency.

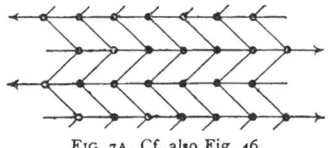

FIG. 7A. Cf. also Fig. 46.

18. The only thing needful to guard one's self
against in following out this mechanical analogy is
the idea that there need be any mechanical displacement of the atoms of matter accompanying
the electric displacement. Manifestly a model containing fixed beams for the attachment of the beads
cannot really correspond very closely to electrical
facts. To make the model imitate facts more closely
we should have to take a number of rows of beads,
each row threaded on its cord and attached crosswise
by elastic threads, as in Fig. 7A.

If these cords are simultaneously displaced alter-

nately in opposite directions, and if they be considered as representing positive and negative electricity
alternately, while the beads represent the electro-positive and the electro-negative elements of the material
substance, then perhaps something more like the
actual state of things may be imagined. There is
here no displacement of the molecule as a whole,
but there is a displacement of its constituent atoms ;
there is a shearing stress applied to each molecule,
which, if strong enough, may result in electrolytic
disruption.   I certainly regard disruptive discharge
as being of this electrolytic character (§ 112).

19. Return, however, to the simple discharge, and see
how it occurs.   Will it take place as a simple sliding
back of the beads to their old position ? Yes, if the
resistance of the circuit is great, but not otherwise.
If the cord is fairly free the beads will fly past their
mean position, overshooting their mark, then rebound, and so, after many quick oscillations, will
finally settle down in their natural position.   Thus
is represented the fact that the discharge of a Leyden
jar is in general oscillatory ; the apparently single and
momentary spark, when analyzed in a very rapidly
rotating mirror, turning out to really consist of a
series of alternating flashes rapidly succeeding one
another, and all over in the hundred-thousandth of a
second or thereabouts.   These oscillatory currents
were predicted and calculated beforehand by Sir

William Thomson ; they were first observed experimentally by Feddersen.[1]   The oscillations continue
until the energy stored up in the strained medium
has all rubbed itself down into heat.   The existence
of these oscillations proves distinctly that electricity
in conjunction with matter possesses inertia.   The
rapidity of these oscillations is something tremendous:
it may reach as high as a hundred million per second,
or it may be as slow as ten thousand per second, according to the capacity and inertia of the circuit.

The rapidity of oscillation, and its rate of dying
away, as well as the circumstances which change the
recovery from an oscillatory one into a simple one-
directioned leak, are precisely analogous to those
which regulate the recovery of a bent spring suddenly
let go.   If the spring is in a very viscous medium, or
if it has but small inertia, it will not oscillate, but will
merely return to its natural position.   Under ordinary
circumstances, however, it will make many oscillations
before its energy is all rubbed into heat (§§ 123 *et seq.*).

Fig. 8 shows part of an actual model of the kind.

20. To make the model represent *charge by induction*
all that has to be done is to immerse a conductor into
the polarized dielectric—in other words, to make one
or more of the beads of the fixed and slippery conducting kind, the other beads on the cord being of

---

[1] We now find that they were experimentally discovered with considerable clearness by Joseph Henry of Washington.   (See p. 369.)

the elastic and adhesive or insulating kind. Then, when the displacement occurs, it is plain that a deficiency of cord will exist on one side of the metallic layer and a surplus on the other, as shown in Fig. 9.

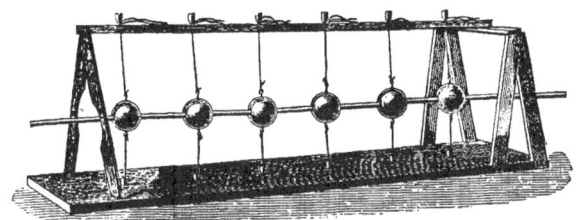

FIG. 8.—Partial model of a dielectric.

This state of things corresponds exactly to the equal opposite induced charges on a conductor under induction, as in Fig. 3.

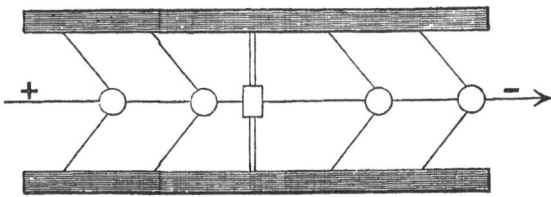

FIG. 9.—Metallic layer in the midst of a polarized dielectric, showing opposite charges "induced" on its surfaces. (Compare Fig. 3.)

If the strain on one side be relieved by letting the beads on that side slip back on the cord, that corresponds to touching the conductor to earth, as in Fig. 4. The other side has now to withstand the whole

E.M.F., consequently the strain there and the charge there will have increased. Remove now the applied E.M.F., and the negative charge appears on both sides of the metal partition, either equally, or more markedly on that side which has fewest beads, *i.e.* which is nearest to other conductors.

21. This being a matter which it is desirable thoroughly to understand, a series of figures illustrating the various stages of the process are appended, in Fig. 9A.

I. represents an ordinary polarized dielectric, say air, between two oppositely charged bodies, A and B, maintained at constant difference of potential. For simplicity the field is taken of uniform strength, *i.e.* with its lines of force parallel straight lines, so that A may be considered as a large positively charged plate, and B an earth-plate facing it. The difference of potential between A and B is called 60, and is distributed among 8 strata or units of thickness, each of which therefore bears the strain $7\frac{1}{2}$, and is displaced $\frac{3}{4}$ of the width of a square from its normal position. The charges on A and B we may call $\pm$ 3 respectively.

An insulated metal plate two units in thickness is now introduced, replacing a couple of the dielectric strata. The remaining 6 have therefore more strain thrown upon them, viz. 10 on each, and accordingly each is now displaced a whole square-width from its normal position, the charge of the plates A and B has

risen to ± 4, and the effect is the same as if the thickness
of dielectric had been reduced in the ratio of 4 to 3·

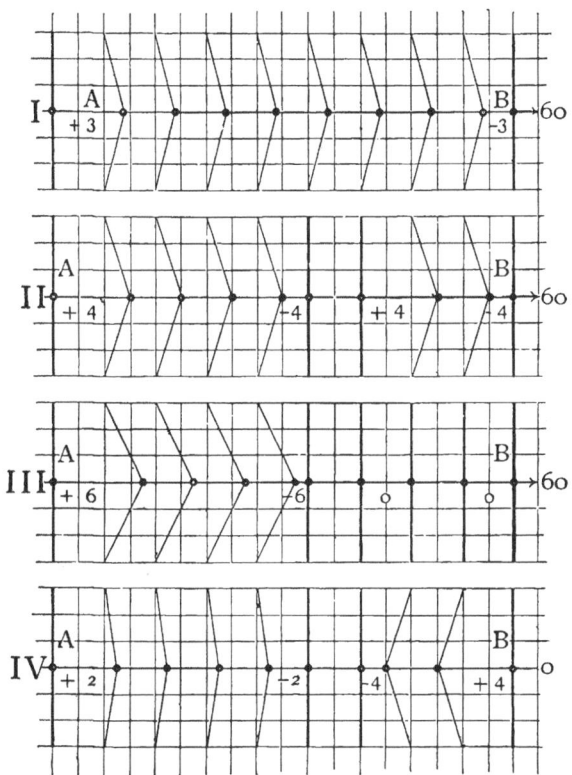

FIG 9A.—Stages during the charge of a metal by induction and contact. Numbers
preceded by + or − represent charges ; numbers affixed to an arrow-head
represent E.M.F.   In this series the E.M.F. applied is supposed constant.

The metal partition introduced has also a charge on

its surface, viz. -4 on the side facing A, and $+4$ on the side facing B. See Diagram II.

The next stage is to connect the metal momentarily to earth. The effect of this is to entirely relieve the strain on the B side by replacing the dielectric with metal, which allows the cord to freely slip through. The cord makes another bound forward, and all the strain is now thrown upon 4 strata, which each have to bear 15, and are displaced $1\frac{1}{2}$ from their natural position. Restoring the dielectric (*i.e.* removing the temporary earth connection) makes no further change, but leaves everything as shown in Diagram III. The charge on one side of the metal partition is now -6, and on the other side is nothing.

Finally remove the constant E.M.F. which has been acting all this time. The cord makes a bound back, its strain becomes nothing ; the 2 strata on the right have to balance the 4 strata on the left, and accordingly their displacements are 1 and $\frac{1}{2}$ respectively. The charges on the faces of the partition are -2 and -4 ; both negative. The charges on A and B are $+2$ and $+4$ respectively, although they are at the same potential. The state of things is shown in IV., and the metal partition has been charged negatively by means of induction. Of course it may have been charged equally on its two faces ; that is a mere matter of the relative proximity of adjacent objects, A and B.

If instead of maintaining A at constant potential

it possessed a constant charge, the series of operations would differ in a slight and easily appreciated manner. The resultant tension on the cord would then be zero all the time, and the series of operations would be practically the electrophorus series, such as go on rapidly and continuously in all inductive machines and replenishers. It will, however, be worth while to sketch this electrophorus series more particularly ; the process of working out what is happening in any given case will then be sufficiently illustrated.

Electrophorus.

22. Diagram I., Fig. 9B, shows the cake excited negatively, resting on its sole. The negative charge on surface of cake is called 13 units ; of these, 12 are what is sometimes called " bound " by the sole, and 1 is free. In other words, the strain due to 12 units of charge is thrown on the layers of the cake, the remaining small strain is thrown on the atmosphere above. The strain in the atmosphere is small because it is so much thicker than the cake—there are so many layers in it that a very small displacement of each suffices to balance the stress in the cord. One unit of charge is induced on the ceiling and walls of the room by the electrified cake. We now bring down the insulated metal cover of the electrophorus. If it is any appreciable thickness it

displaces a few of the strained layers, and thus there
is a little extra strain on the others ; but this effect is

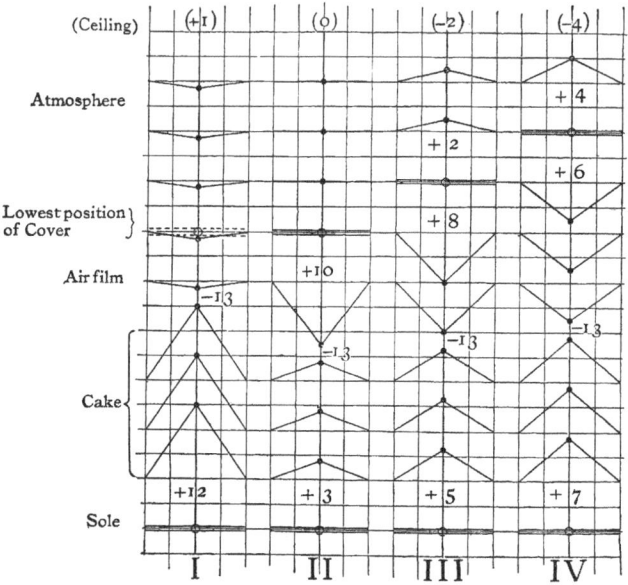

Fig. 9B.—The electrophorus series of operations.   Double lines mean rigid rods
supporting smooth beads, and represent metal.   Gripping beads supported by
elastics represent dielectric.   Numbers preceded by + or − represent charges.
The charge excited on the surface of the electrophorus cake remains constant all
the time.   The sole is supposed connected with the floor and walls of the room.
Numbers in parentheses at the top represent the charge induced on ceiling and
walls.   The whole thickness of atmosphere does not pretend to be represented.
It must be thick enough in I. precisely to balance the stress in the cake.   The
resultant stress in the cord is in each case zero.

I. shows the charged cake, with cover either off or in position but insulated.
II. shows effect of connecting cover and sole.
III. and IV. show effect of gradually raising cover.
I. again shows effect of having removed cover and discharged it.

extremely small, and it is quite unessential.   We
may therefore take the cover as of no thickness, and

bring it down into what is marked in the diagram as its lowest position; the stress passes through it, and nothing is affected except the one layer whose place it takes. Diagram I. will serve to represent the cover thus put on, so long as it is insulated. The dotted lines show it in position. It does not make intimate contact with the cake ; a film, either of air or of the substance of the cake itself, intervenes between it and the negatively charged surface, and this is exhibited in the diagram.

The next thing is to connect the cover and the sole together.   This immediately brings about the state of things represented by Diagram II.

A charge of 9 units has rushed round from sole to cover, making with the charge 1 which previously existed on the walls of the room a total of 10.[1] The strain above the cover is entirely relieved, and the whole excitement is now internal between cover and sole.   The strain in the cake is considerably relieved, but the work of balancing what remains in it is thrown on the very thin film between cover and top of cake.   This, therefore, is highly strained.

We now raise the again insulated cover. As it ascends fresh layers of dielectric intervene between it and the cake, and receive some of the strain.   The effect

---

[1] If the sole had been insulated, and connection between it and cover also made in an insulated manner, then this unit on the walls of the room would stay there ; the cover would only acquire a charge 9, and the slight strain above it shown in I. would continue to exist unaltered.

E

of this is threefold. First, they partially relieve the
strain in the original very thin layer ; next, they
increase the strain in the cake ; and thirdly, they
put a little strain on the air above the raised cover.
The sole therefore receives 5 units instead of 3 ; the
cover retains its charge 10, but part of this is on its
upper surface ; the induced charge − 2 makes its
appearance on the walls of the room. The state
is shown in Diagram III.

Diagram IV. continues the process of raising,
until ultimately when the plate is removed to
infinity, its charge above and below is equal,
being 5 on each, and the cake and cover have
returned to their original state I., ready to begin
again. The cover having now a charge 10, the
walls of the room wherever it is will have a charge
− 10, and it may be discharged whenever one pleases
without affecting the cake at all. Having discharged
it, one can put it on, as in I., and perform the
cycle again.

If one chooses to put the cover on before discharging it, the cycle of operations is just reversed,
from IV. to II.

It is instructive to mount an electrophorus on
an insulating stand and connect its sole to earth
through a delicate galvanometer ; then the rush out
of it when the cover is touched, and the flow back
again as the cover is raised, can be easily watched.

23. There is one more thing which is so important to see clearly that an illustration of it is desirable, and that is the effect of inserting not a metal but a slice of some other perfectly insulating dielectric, with a different inductive capacity, in the midst of a polarized medium. Thus, for instance, between the plate of a charged condenser insert a thick slab of glass. The effects will differ according as the condenser plates are charged each with a given quantity, or are maintained at a constant difference of potential.

Refer to Fig. 9C ; the 8 similar strata are supposed to be displaced with a total E.M.F. 24, the tension in the cord (negative electric potential) accordingly rises by a step 3 at each layer. Diagram I. shows this initial state. Clamp the cord, to represent a constant charge on the plates A and B, and now introduce a slab of glass—that is, replace the 4 middle layers by elastics only half as stiff (see II.). The stress in the cord steps up now by only $1\frac{1}{2}$ at each of these layers, and the total difference of potential, instead of being 24, is now only 18. Meanwhile the charges remain the same, and there is no charge on the surface of the glass ; the capacity of the whole condenser has increased in the ratio of 4 to 3.

There is no charge on the surface of the glass ; but the resultant effect is very much the same as if there were. The effect on the cord will be precisely the same as if the replaced elastics were still of the same

E 2

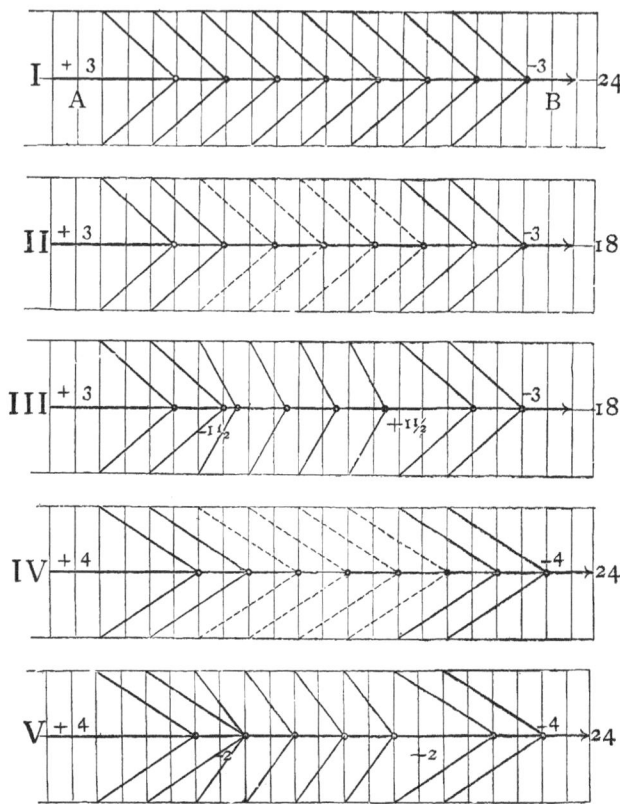

FIG. 90 —Real and apparent effects of introducing a glass slab between the plates of an air-condenser.

I. shows the original condenser, of capacity ¼.
II. shows the effect of inserting a slab of half the whole thickness, and of specific inductive capacity 2, the charge being kept constant. The capacity has risen to ⅓.
III. shows a spurious imitative mode of obtaining the same effect, without any change of inductive capacity, by help of surface charges.
IV. shows the effect of introducing the slab into the condenser when it is supplied with a constant E.M.F. The capacity has again become ⅓.
V. shows a spurious imitation of this effect by help of surface charges.

strength, but as if their beads had slid half-way back, into the positions shown in III., where surface charges exist as indicated by numbers. This, I repeat, is *not* the state of things caused by the glass, but it is so like it in effect as to be difficultly distinguishable from it ; and one sometimes speaks of the spurious or virtual charges set up on the glass surface, meaning the charges in Diagram III., which so exactly imitates the resultant effect of II.

So much for the effect of constant charge ; now take the case of constant potential.

Diagram IV. shows the effect of replacing some of the elastics by weaker ones in this case.   The E.M.F. is kept constant, so the strong elastics have more strain thrown on them than before ; no internal charge is possible so long as the substances insulate perfectly, so all the beads are pulled forward equally. The step of potential is now 4 at all the stiffer (or air) strata, and 2 at all the weaker (or glass) strata, making up the total E.M.F., 24.   The charge on the plates A and B has increased from $\pm$ 3 to $\pm$ 4 in accordance with the increase of capacity, the rate of increase of which is still 4 : 3.   Here again the real effect shown in IV. may be simulated by spurious surface charges without any change of inductive capacity, as is sufficiently indicated by Diagram V., wherein all the elastics are supposed of the same strength.

24. *Hydraulic Model of a Leyden Jar.*—So much
for the cord model, but I will now describe and explain an hydraulic model which illustrates the same
sort of facts ; some of them more plainly and directly
than the cord model. Moreover, since all charging
is essentially analogous to that of a Leyden jar,
let us take a Leyden jar and make its hydrostatic
analogue at once.

The form of jar most convenient to think of is one
supported horizontally on an insulating stand, with
pith ball electroscopes supplied to both inner and outer
coatings. Or one may use, as I commonly do, in conjunction with the hydraulic model, a vertical coated
pane, with a pith ball connected to each coating ; but
if the electroscopes were of such a kind as to show a
difference between positive and negative potential,
they would do better.

To construct its hydraulic model, procure a thin
india-rubber bag, such as are distended with gas at
toy-shops ; tie it over the mouth of a tube with
a stop-cock A, and insert the tube by means of a
cork into a three-necked globular glass vessel or
" receiver," as shown in the diagram, Fig. 10.

One of the other openings is to have another
stop-cock tube, B ; and the third opening is to be
plugged with a cork as soon as the whole vessel, both
inside and outside the bag, is completely full of water
without air-bubbles.

This is the insulated Leyden jar : the bag represents the dielectric, and its inner and outer coatings are the spaces full of water.

Open gauge-tubes, a and b, must now be inserted in tubes A and B, to correspond to the electroscopes supplied to the jar ; and a third bent tube, C, con-

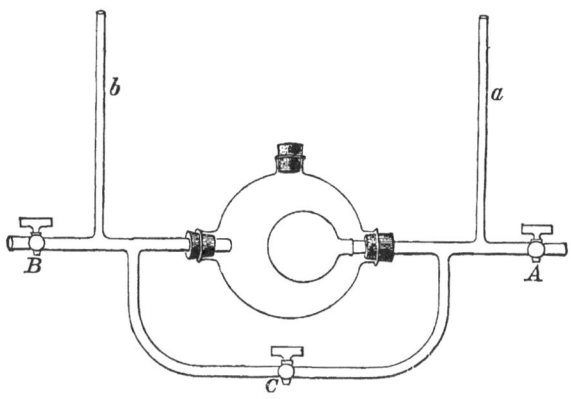

FIG. 10.—Skeleton diagram of hydraulic model of a Leyden jar.

necting the inner and outer coatings, will correspond to a discharger. Ordinarily, however, of course C will be shut.

A water-pump screwed on to A will represent an electric machine connected to inner coating ; and the outer coating, B, should open into a tank, to represent the earth. The pump will naturally draw its supply of water from the same tank.

The bag being undistended, and the whole filled with water free from air, the level of the water in the two gauge-tubes will correspond with that in the tank ; and this means that everything is at zero potential, *i.e.* the potential of the earth.

Now, C being shut, shut also B, open A, and work the pump. Instantly the level in the two gauges rises greatly and equally : you are trying to charge an insulated jar. Turn an electric machine connected to a real jar, and its two pith balls will similarly and equally rise.

Now open B for an instant, the pressure is relieved, and both gauges at once fall, apparently both to zero. Repeat the whole operation several times however, and it will be found that, whereas *b* always falls to zero, *a* falls short of zero each time by a larger amount, and the bag is gradually becoming distended. This is *charge by alternate contact.* It may be repeated exactly with the real jar : a spark put into the inner coating, and an equal spark withdrawn from the outer coating each time ; and unless this outer spark is so withdrawn, the jar declines to charge : water (and electricity) being incompressible.

If B is left permanently open, the pump can be steadily worked, so as to distend the bag and raise the gauge *a* to its full height, *b* remaining at zero all the time, save for oscillatory disturbances.

Having got the jar charged, shut A, and remove

the pump, connecting the end of A with the tank directly.

Now of course by the use of the discharger C the fluid can be transferred from inner to outer coat, the strain relieved, and the gauges equalized. But if this operation be performed while the jar is insulated, *i.e.* while A and B are both shut, the common level of the gauges after discharge is not zero, but a half-way level ; and the effect of this is very noticeable if you touch an insulated Leyden jar after it has been discharged.

Instead of using the discharger C, however, we can proceed to discharge by alternate contact, and the operation is very instructive.

Start with the gauge *b* at zero, and the gauge *a* at high pressure. Open stop-cock A ; some water is squeezed out of inner coating, and the *a* gauge falls to zero, but the suck of the contracting bag on the outer coat pulls down the gauge *b below* zero, the descent of the two gauges being nearly equal.

Next shut A and open B ; a little water flows in from the tank to still further relieve the strain of the bag, and both gauges rise ; *b* to zero, *a* to something just short of its old position.

Now shut B and open A again : again the two gauges descend. Reverse the taps, and again they both rise ; and so on until the bag has recovered its normal size. This is discharge by alternate contact, and exactly

imitates the behaviour of an insulated charged Leyden
jar whose inner and outer coats are alternately touched
to earth.   Its pith balls alternately rise with positive
and with negative electricity, indicating potentials
above and below zero.

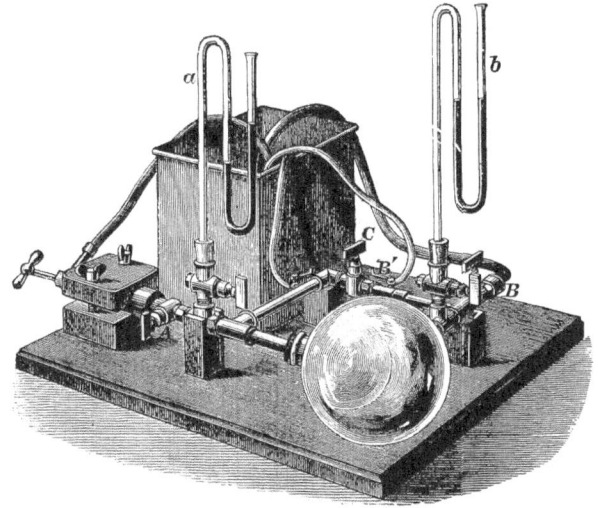

Fig. 11.—First actually constructed model Leyden jar, with mercury gauges or
    electrometers ; the whole rigged up with things purchasable at a plumber's,
    except the pump. The glass globe contains an elastic bag, which swells as
    water is pumped into it. The tank is kept full of water, and its level represents
    the potential of the earth. Flexible tubes full of water effect the desired
    earth-connections when wished.   The gauges *a* and *b* represent electroscopes
    connected to inner and outer coats of the jar respectively.

Figs. 11 and 12 are taken from photographs of
apparatus I have made to use as just described.   The
glass globe with the partially distended bag inside it,
the pump, the tank, the gauges *a* and *b*, the stop-cocks

A B C, will be easily recognized.   Two extra stop-cocks,
A' and B', leading direct to tank, are extra, and are to

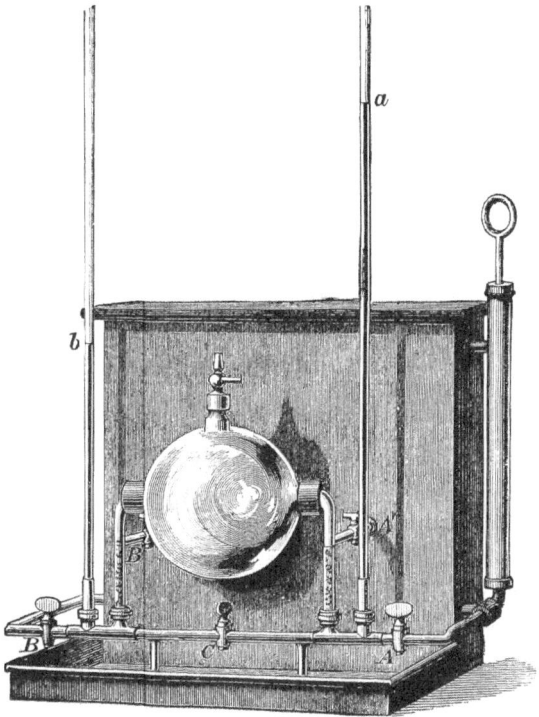

FIG. 12.—Latest form of hydraulic model of a Leyden jar with water gauges ; the
whole arranged vertically to be more conspicuous.   The pump is a force-pump
with a communication between top of barrel and tank to get rid of stray water.
The parts are labelled to correspond with the skeleton diagram, Fig. 10, as well
as with Fig. 11.

save having to disconnect pump and connect A direct,
when exhibiting the effect of " discharge by alternate

contact." But the tank is not sufficiently tall in Fig. 12 ; I have doubled its height since. The full height of the gauge-tubes is barely shown.

In any form of apparatus it is essential to fill the whole with water—pipes, globe, everything—before commencing to draw any moral from its behaviour. It is rather difficult to get rid of a large bubble of air from the top of the globe of Fig. 11, and though it is not of very much consequence in this place, the stop-cock in Fig. 12 is added to make its removal easy. The gauges in Fig. 11 may be replaced by others arranged as a lantern-slide, and connected by flexible tubing full of air.

25. I have explained thus fully the hydraulic illustration of Leyden jar phenomena, because these constitute the key to a great part of electrostatics. The illustration is not indeed a complete or perfect one by any means, but by combining with it a consideration of the endless cord models, and of what I have endeavoured to explain concerning conduction and insulation in general, a distinct step may be gained.

Think of electrical phenomena as produced by an all-permeating liquid embedded in a jelly ; think of conductors as holes and pipes in this jelly, of an electrical machine as a pump, of charge as excess or defect, of attraction as due to strain, of discharge as bursting, of the discharge of a Leyden jar as a springing back or recoil, oscillating till its energy has gone.

By thus thinking you will get a more real grasp of the subject and insight into the actual processes occurring in Nature—unknown though these may still strictly be—than if you employed the old ideas of action at a distance, or contented yourselves with no theory at all on which to link the facts. You will have made a step in the direction of the truth, but I must beg you to understand that it is only a step ; that what modifications and additions will have to be made to it before it becomes a complete theory of electricity I am unable fully to tell you. I am convinced they will be many, but I am also convinced that it is unwise to drift along among a host of complicated phenomena without guide other than that afforded by hard and rigid mathematical equations.

The mathematical theory of potential and the like has insured safe and certain progress, and enables mathematicians to dispense for the time being with theories of electricity and with mental imagery. Few, however, are the minds strong enough thus to dispense with all but the most formal and severe of mental aids ; and none, I believe, to whom some mental picture of the actual processes would not be a help if it were safely available.

Such a representation I have endeavoured partially to lay before you; and I hope, if I have succeeded in making myself at all intelligible, that students of electricity will find it of some use and service.

PART II.

CONDUCTION.

CHAPTER IV.

METALLIC AND ELECTROLYTIC CONDUCTION.

26. WE have now glanced through electrostatic phenomena, and seen that they could be all comprehended and partially explained by supposing electricity to be a fluid of perfect incompressibility—in other words, a liquid—permeating everywhere and everything ; and by further supposing that in conducting matter this liquid was capable of free locomotion, but that in insulators and general space it was as it were entangled in some elastic medium or jelly, to strains in which electrostatic actions are due. This medium might be burst, in a disruptive discharge, but easy flow could go on only through channels or holes in it, which therefore were taken to represent conductors ; and it was obvious that all flow must take place in closed circuits.

To-day I want to consider the circumstances of this flow more particularly : to study in fact, the second

F

division of our subject (see classification on page 4), viz. *Electricity in locomotion.*

I use the term "locomotion" in order to eliminate rotation and vibration: it is translation only with which we intend now to concern ourselves.

Consider the modes in which *water* may be made to move from place to place ; there are only two : it may be pumped along pipes, or it may be carried about in jugs. In other words, it may travel *through* matter, or it may travel *with* matter. Just so it is with *heat* also ; heat can travel in two ways : it can flow *through* matter, by what is called "conduction," and it can travel *with* matter, by what is called "convection." There is no other mode of conveyance of heat. You frequently find it stated that there is a third method, viz. "radiation"; but this is not truly a conveyance of *heat* at all. Heat generates radiation at one place, and radiation reproduces heat at another ; but it is radiation which travels, and not heat. Heat only naturally flows from hot bodies to cold, just as water only naturally flows down hill ; but radiation spreads in all directions, without the least attention to where it is going. Heat can only flow one way at any given point, but radiation travels all ways at once. If water were dissociated on one planet into its constituent gases, and if these recombined on another planet, it would not be water which travelled from one to the other, neither would the substance obey the laws of motion of water—water would be destroyed in one place, and reproduced in another; just so is it with the relation between radiation and heat.

Heat, then, like water, has but two direct modes of conveyance from place to place. For *electricity* the same is true. Electricity can travel with matter, or it can travel through matter; by convection or by conduction, but in no other known way.

Conduction in Metals.

27. Consider, first, conduction. Connect the poles of a voltaic battery to the two ends of a copper wire, and think of what we call the " current." It is a true flow of electricity among the molecules of the wire. If electricity were a fluid, then it would be a transport of that fluid ; if electricity is nothing material, then a current is no material transfer ; but it is certainly a transfer of electricity, whatever electricity may be. Permitting ourselves again the analogy of a liquid, we can picture it flowing through, or among, the molecules of the metal. Does it flow through or between them ? Or does it get handed on from one to the next continually? We do not quite know ; but the last supposition is often believed to most nearly represent the probable truth. The flow may be thought of as a perpetual attempt to set up a strain like that in a

dielectric, combined with an equally perpetual breaking down of every trace of that strain. If the atoms be conceived as little conductors vibrating about and knocking each other, so as to be easily and completely able to pass on any electric charge they may possess, then, through a medium so constituted, electric conduction could go on much as it does go on in a metal. Each atom would receive a charge from those behind it, and hand it on to those in front of it, and thus may electricity get conveyed along the wire. Do not, however, accept this picture as anything better than a *possible* mode of reducing conduction to a kind of electrostatics—an interchange of electric charges among a series of conductors. If such a series of vibrating and colliding particles existed, then certainly a charge given to any point would rapidly distribute itself over the whole, and the potential would quickly become uniform ; but it by no means follows that the actual process of conduction is anything like this. Certainly it is not the simplest mode of picturing it for ordinary purposes. The easiest and crudest idea is to liken a wire conveying electricity to a pipe full of marbles or sand conveying water ; and for many purposes, though not for all, this crude idea suffices.

Leaving the actual mode of conveyance as unknown, let us review how much is certainly known of the process called conduction in homogeneous metals.

This much is certainly known :—

(1) That the wire gets heated by the passage of a current.

(2) That no trace of a tendency to reverse discharge or spring back exists.

(3) That the electricity meets with a certain amount of resistance or friction-like obstruction.

(4) That this force of obstruction is accurately proportional to the speed with which the electricity travels through the metal—that is, to the intensity of the current per unit area.

28. About this last fact a word or two must be said. The amount of electricity conveyed per second across a unit area is called the intensity of current,[1] and experiment proves, what Ohm originally guessed as probable from the analogy of heat conduction, that this intensity is accurately proportional to the slope of potential which causes the flow ; or, in other words (since action and reaction are equal and opposite), that a current in a conductor meets with an obstructive electromotive force exactly proportional to itself. Or, quite briefly, a current through a given conductor is proportional to the E.M.F. which drives it. The particular ratio between slope of potential and corresponding intensity of current depends upon the particular material of which the conductor is composed,

---

[1] Often called "density" of current, but "intensity" is the natural and proper expression for the purpose.

and is one of the constants of the material, to be determined by direct measurement. It is called its "specific conductivity" or its "specific resistance" according to the way it is regarded.

The law here stated is called Ohm's law, and is one of the most accurately known laws there are. Nevertheless it is an empirical relation ; in other words, it has not yet been accounted for—it must be accepted as an experimental fact. Undoubtedly, it is one of vast and far-reaching importance : it asserts a connection between electricity and ordinary matter of a definite and simple kind. Using the language of hydraulic analogy, it asserts that when electricity flows through matter the friction between them is accurately as the first power of the velocity for all speeds.

29. Now if we think of this opposing electromotive force as analogous to friction, it is very easy to think of heat being generated by the passage of a current, and to suppose that the rate of heat-production will be directly proportional to the opposing force and to the current driven against it—as in fact Joule experimentally proved it to be.

But if we are not satisfied with this vague analogy, and wish to penetrate into the ultimate nature of heat and the mode in which it can be generated, then we can return to the consideration of a multitude of oscillating and colliding particles, moving with a certain average energy which determines what we call the

" temperature" of the body. If now one or more of
these bodies receives a knock, the energy of the blow
is speedily shared among all the others, and they all
begin to move rather more energetically than before :
the body which the assemblage of particles constitutes
is said to have " risen in temperature." This illustrates
the production of heat by a blow or other mechanical
means. But now, instead of *striking* one of the balls
give it an electric charge ; or, better still, put within its
reach a constant reservoir of electricity from which it
can receive a charge every time it strikes it, and at
the same time put within the reach of some other of
the assemblage of particles another reservoir of infinite capacity which shall be able to drain away all the
electricity it may receive. In practice there is no
need of infinite reservoirs : all that is wanted is to
connect two finite reservoirs, or "electrodes," as one
might now call them, with some constant means of
propelling electricity from one to the other, *i.e.* with
the poles of a voltaic battery or a Holtz machine.

What will be the result of thus passing a series of
electric charges through the assemblage of particles ?
Plainly the act of receiving a charge and passing it on
will tend to increase the original motion of each particle ; it will tend to raise the temperature of the
body. In this way, therefore, it is possible to picture
the mode in which an electric current generates heat.

But although this process may be used as a possible

analogy, it cannot be a true and complete statement of
what occurs ; for it is essentially the mode of propagation of *sound*. Sound travels at a definite and known
velocity, being a mechanical disturbance handed on
from particle to particle in the manner described.
But heat, being some mode of motion, must also be
handed on after some analogous fashion, so that when
heat is supplied to one point of a mass it spreads or
diffuses through it. It is difficult to suppose the conduction of heat to be other than the handing on of
molecular quiverings from one to another, and yet it
takes place according to laws altogether different from
those of the propagation of the gross disturbance
called sound. The exact mode of conduction of heat is
unknown, but, whatever it is, it can hardly be doubted
that the conduction of electricity through metals is
not very unlike it, for the two processes obey the same
laws of propagation : they are both of the nature of a
diffusion, they both obey Ohm's law, and a metal
which conducts heat well conducts electricity well
also.

Conduction in Liquids.

30. Leaving the obscure subject of conduction in
metals for the present, let us pass to the consideration
of the way in which electricity flows through liquids.
By " liquids," in the present connection, one more par-

ticularly means definite chemical compounds, such as acids, alkalies, salt and water, and saline solutions generally. Some liquids there are, like alcohol, turpentine, bisulphide of carbon, and possibly water, which, when quite pure, either wholly or very nearly decline to conduct electricity at all. Such liquids as these may be classed along with air and gases as more or less perfect dielectrics. Other liquids there are, like mercury and molten metals generally, which conduct after precisely the same fashion as they do when solid. These, therefore, are properly classed among metallic conductors.

But most chemical compounds, when liquefied either by heat or by solution, conduct in a way peculiarly their own ; and these are called " electrolytes."

31. The present state of our knowledge enables us to make the following assertions with considerable confidence of their truth :—

(1) Electrolytic conduction is invariably accompanied by chemical decomposition, and in fact only occurs by means of it.

(2) The electricity does not flow *through*, but *with*, the atoms of matter, which travel along and convey their charges something after the manner of pith balls.

(3) The electric charge belonging to each atom of matter is a simple multiple of a definite quantity of

electricity, which quantity is an absolute constant quite independent of the nature of the particular substance to which the atoms belong.

(4) Positive electricity is conveyed through a liquid by something equivalent to a procession of the electro-positive atoms of the compound, in the direction called the direction of the current ; and at the same time negative electricity is conveyed in the opposite direction by a similar procession of the electro-negative atoms.

(5) On any atom reaching an electrode it may be forced to get rid of its electric charge, and, combining with others of the same kind, escape in the free state ; in which case visible decomposition results. Or it may find something else handy with which to combine—say on the electrode or in the solution ; and in that case the decomposition, though real, is masked, and not apparent.

(6) But, on the other hand, the atom may cling to its electric charge with such tenacity as to stop the current : the opposition force exerted by these atoms upon the current being called polarization.

(7) No such opposition force, or tendency to spring back, is experienced in the interior of a mass of fluid : it occurs only at the electrodes.

32. The first three of these statements constitute a summary of Faraday's laws of electrolysis. These laws are of far-reaching importance, and appear to be

accurately true. The first is called the "voltametric law," and asserts that the amount of chemical action electrolytically produced in any given substance is exactly proportional to the quantity of electricity that has passed through it. The vague phrase " chemical action " is purposely used here to include decomposition or recomposition or liberation or deposition or dissolution, or any other effect that can be brought about in either elements or compounds by the passage of an electric current. The weight of substance acted on measures the quantity of electricity which has passed ; hence a decomposition cell can act as a voltameter, and the law is called the voltametric law. Its truth enables us to make the first of the above statements ; which many qualitative facts concerning the details of electrolysis modify into statement No. 2.

The second of Faraday's laws is called the law of electro-chemical equivalence, and asserts that, if the same current be passed through a series of voltameters for the same time, the amount of chemical action in each substance acted on is exactly proportional to its ordinary chemical equivalent ; not to its atomic weight merely, but to its atomic weight divided by what is called its valency, or atomicity, or quantivalence ; this being its real chemical equivalent. Thus an atom of oxygen weighs 16 times as much as an atom of hydrogen, and is equivalent to two such atoms in combining power ; hence the law asserts that

8 grammes of oxygen are liberated for every gramme of hydrogen. Again, an atom of silver is 108 times as heavy as an atom of hydrogen, and is equal to it in combining power ; hence 108 grammes of silver are deposited in a silver voltameter while 1 gramme of hydrogen is being liberated by the same current in a gas voltameter. Once more, an atom of gold weighs as much as 197 atoms of hydrogen, and is able to replace three of them in combination ; hence 65·7 grammes of gold are deposited by the same current in the same time, and so on.

Now this law plainly means that the same number of monad atoms is liberated by the same quantity of electricity no matter what their nature may be ; half that number of dyad atoms ; one third that number of triad atoms. Hence, assuming statement (2), that the current flows by convection—each atom conveying electricity—it follows that every monad atom carries the same quantity, whether it be an atom of hydrogen or of silver or of chlorine, or a complex radicle like NO_3 ; that each dyad atom carries twice as much, whether it be an atom of oxygen or of zinc or of copper, or a complex dyad radicle like SO_4 ; that each triad atom carries three times as much, and so on. And this is what is laid down in the third of the above statements.

True, it is possible that every atom may have a specific charge of its own with which it never parts ;

but about such nothing is known ; we can only make experiments on the charge it is willing to part with at an electrode, and there is no doubt that this is accurately the same for all substances, up to a simple multiple. And this quantity, the charge of one monad atom, constitutes the smallest known portion of electricity and is a real natural unit. Obviously this is a most vital fact. This unit below which nothing is known has even been styled an "atom" of electricity ; and perhaps the phrase may have some meaning. I have ventured to suggest one or two effects which would result from the hypothesis that this unit quantity of electricity were really in fact an absolute minimum, and as indivisible as an atom of matter.[1] The natural unit of electricity is exceedingly small, being about the hundred-thousand-millionth part of the ordinary electrostatic unit ; or less than the hundred-trillionth of a coulomb.

The charge of each atom being so small, its potential is not high. Something between 1 and 3 volts is a probable difference of potential for two oppositely charged atoms. But they are so near together that even this small difference of potential causes a strong electrostatic attraction or "chemical affinity" between the oppositely charged atoms.

This electrical force between two atoms at any

---

[1] See paper on " Electrolysis " at Aberdeen (*Reports of the British Association for* 1885, p. 763).

distance is ten thousand million billion billion times
greater than their gravitative attraction at the same
distance. The force has an intensity per unit mass
(and therefore is able to produce an acceleration) nearly
a trillion times greater than that of terrestrial gravity
near the earth's surface.

These are undoubtedly the forces with which
chemists have to do, and which they have long called
chemical affinity.

33. But it may be asked, If the atoms in each
molecule cling together by their electrostatic attractions,
and there are an enormous number of atoms between
two electrodes, how comes it that a feeble E.M.F. can
pull them apart and effect decomposition ; moreover,
how can the E.M.F. needed to effect decomposition help
varying directly with the thickness of fluid between
the plates ? It does not depend on anything of the
kind ; the length of liquid between the electrodes is
absolutely immaterial. This proves that throughout
the main thickness of liquid no atoms are torn asunder
at all. Probably they frequently change partners, one
pair of atoms not always remaining united but occa-
sionally getting separated and recombined with other
individuals. During these interchanges there must be
moments of semi-freedom during which the atoms are
amenable to the slightest directive tendency, and it is
probably these moments that the applied E.M.F. makes
use of.

The reality of such a state of continual interchange between molecules has been forced upon chemists by the facts of double-decomposition—such facts as the interchange of atoms between strongly combined salts when their solutions are mixed so as to form very much weaker compounds ; the proof that such compounds are formed being very clear in the case when they happen to be insoluble. The fact that if a precipitate is insoluble enough it is bound to form, really proves that some small quantity of the corresponding compound is always formed in every case, whether it happens to be insoluble or not.

The state of continual interchange results in a perfect sensibility to the migratory tendency of extremely weak forces, so that even the faintest trace of an electromotive force is able to affect the charged atoms, on the average assisting the positive atoms down the slope of potential, and the negative atoms up the slope. The fact that the most infinitesimal force is sufficient to effect its due quota of decomposition has been proved most clearly and decisively by the experiments of Helmholtz.

Sometimes the term dissociation is used to signify this practical freedom of atoms to locomotion ; and, as stated originally by Prof. Clausius, the idea of dissociation was certainly involved. It was thought that a certain percentage of atoms existed in the liquid in an uncombined state, wandering about

seeking partners, that it was these loose atoms on
which the electromotive forces acted, and that the
procession of these conveyed the current. But we now
see that the addition of the idea of double-decomposition and interchange to the original hypothesis of
Grotthus explains all that is required by the facts, viz.
a virtual or potential dissociation, a momentary state
of hovering and indecision, without the need for any
continuous and actual dissociation.

34. I will now try and make the process of
electrolytic conduction clearer by reverting to our
mechanical analogies and models.

Looking back to Figs. 5 and 6, we see illustrations
of metallic conduction and of dielectric induction.   In
each case an applied electromotive force causes some
m ,vement of electricity ; but, whereas in the first case
it is a continuous almost unresisted movement or
steady flow through or among the atoms of matter, in
the second case it is a momentary shift or displacement only, carrying the atoms of matter with it, and
highly resisted in consequence :—resisted, not with a
mere frictional rub, which retards but does not check
the motion, but by an active spring back force, which
immediately checks all further current, produces what
we call "insulation," and ultimately, when the propelling force is removed, causes a quick reverse motion
or discharge.   But the model is plainly an incomplete
one ; for what is it that the atoms are clinging to ?

What is it ought to take the place of the *beam* in the
crude mechanical contrivance ? Obviously another
set of atoms, which are either kept still or urged in the
opposite direction by a simultaneous opposite displacement of negative electricity ; as in Fig. 7A. We are
to picture two or any number of rows of beads, each
row threaded on its appropriate cord ; the cords
alternately representing positive and negative electricity respectively, and being simultaneously displaced
in opposite directions by any applied E.M.F. The
beads threaded on any one cord have, in a dielectric,
elastic attachments to those on some opposite cord,
and thus continuous motion of the cords in opposite
directions is prevented : only a slight displacement is
permitted, followed by a spring back and oscillation
after the fashion already described.

Very well ; now picture the elastic connections between the beads all dissolved, and once more apply a
force to each cord, moving half of them one way and the
alternate half the other way, and you have a model
illustrating an electrolyte and electrolytic conduction.
The atoms are no longer attached to each other, but
they are attached to the cord. In the first respect, an
electrolyte differs from a dielectric ; in the second, it
differs from a metal.

Moreover, electrolytic conduction is perceived to
be scarcely of the nature of true conduction : the
electricity does not slip through or among the

G

molecules, it goes with them. The constituents of each molecule are free of each other, and while one set of atoms conveys positive electricity, the other set carries negative electricity in the opposite direction ; and so it is by a procession of free atoms that the current is transmitted. The process is of the nature of convection : the atoms act as carriers. Free loco-motion of charged atoms is essential to electrolysis.

35. In order to compare with Figs. 5 and 6, so as to bring out the points of difference, Fig. 13 is drawn. The beads representing one set of atoms of matter are tightly attached to the cord, no trace of slip between them being permitted, but otherwise they are free, and so are represented as supported merely by rings sliding freely on glass rods. The only resistance to the motion, beside the slight friction, is offered at the electrode, which is typified by the spring-backed knife-edge, Z. This is supposed to be able to release the beads from the cord when they are pressed against it with sufficient force. The cling between the bead and cord (*i.e.* between each atom and its charge) is great enough to cause a perceptible compression of the springs, and accordingly to bring out a recoil force in imitation of polarization.

The piece of cord accompanying each bead on its journey (*i.e.* the length between it and the next bead) represents the atomic charge, and is a perfectly con-stant quantity : the only variation permissible in it is that some kinds of atoms have twice as much, or are twice as far apart on their cord, and these are called by chemists dyad atoms ; another kind has three times as much, another four, and so on ; these being called triad, tetrad, &c.

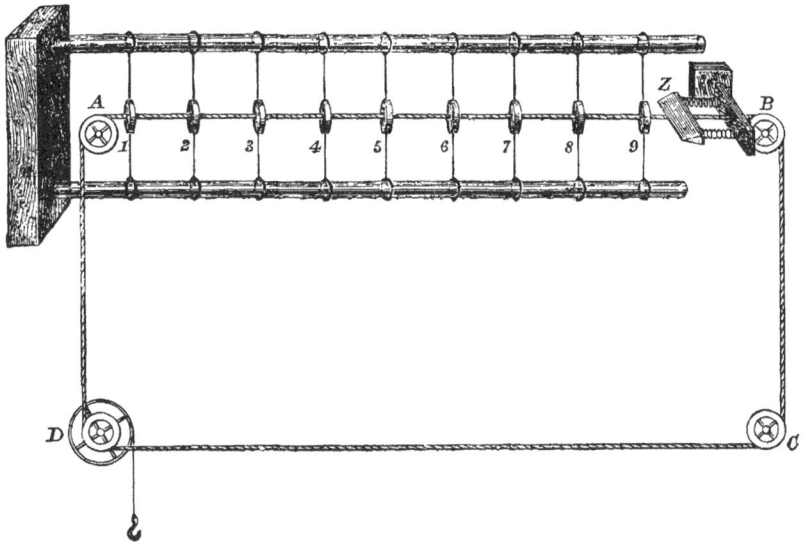

FIG. 13.—Crude mechanical analogy, illustrating a few points in a circuit partly *electrolytic.*

If the cord be taken to represent positive electricity, the beads on it may represent atoms of hydrogen, or other monad *cation*, travelling down stream to the cathode. Another cord representing negative electricity may be ranged alongside it, with its beads twice as

G 2

far apart, to represent the atoms of a dyad *anion*, like oxygen. If the cords are so mechanically connected that they must move with equal pace in opposite directions, we have a model illustrating several important facts. The number of oxygen atoms liberated in a given time will be obviously half the number of hydrogen atoms set free in the same time, and will therefore in the gaseous state occupy but half the volume. For any element whatever, the number of atoms liberated in any time is equal to the number of atoms of hydrogen liberated in the same time, divided by the "valency" of the element as compared with hydrogen. This law was discovered by Faraday, and appears to be precisely true ; and inasmuch as the relative weight of every element is known with fair accuracy, it is easy to calculate what weight of substance any given current will deposit or set free in an hour, if we once determine it experimentally for any one substance.

We may summarize thus :—

If we apply E.M.F. to a metal we get a continuous flow, and the result is heat.

If we apply it to a dielectric we get a momentary flow or displacement, and the result is the potential energy of "charge."

If we apply it to an electrolyte we again get a continuous flow, and the result is chemical decomposition.

36. There are a large number of important points

to which I might direct your attention in the mode by
which an electric current is conveyed through liquids,
but I will specially select one, viz. that it is effected
by a procession of positively charged atoms travelling
one way, and a corresponding procession of negatively
charged atoms the other way.

Whatever we understand by a positive charge and
a negative charge, it is certain that the atoms of, say,
a water molecule, are charged, the hydrogen positively,
the oxygen negatively ; and it is almost certain that
they hang together by reason of the attraction
between their opposite charges. It is also certain
that when an electromotive force—*i.e.* any force
capable of propelling electricity—is brought to
bear on the liquid, the hydrogen atoms travel on
the whole in one direction, viz. down hill, and the
oxygen atoms travel in the other direction, viz. up
hill; using the idea of level as our analogue for
electric potential in this case. The atoms may be
said to be driven along by their electric charges
just as charged pith balls would be driven along ;
and they thus act as conveyers of electricity, which
otherwise would be unable to move through the liquid.

Each of this pair of opposite processions goes on
until it meets with some discontinuity—either some
change of liquid, or some solid conductor. At a
change of liquid another set of atoms continues
the convection, and nothing very particular need be

noticed at the junction ; but at a solid conductor the
stream of atoms must stop : you cannot have locomotion of the atoms of a solid.   The obstruction so
produced may stop the procession, and therefore the
current, altogether ; or on the other hand the force
driving the charges forward may be so great as to
wrench them free, to give the charges up to the electrode which conveys it away by common conduction,
and to crowd the atoms together in such a way that
they are glad to combine with each other and escape.

Now notice the fact of the two opposite processions.
One cannot have a procession of positive atoms
through a liquid without a corresponding procession
of negative ones.   In other words, an electric current
in a liquid necessarily consists of a flow of positive
electricity in one direction, combined with a flow of
negative electricity in the opposite direction.   And if
this is thus proved to occur in a liquid, why should it
not occur everywhere ?   It is at least well to bear the
possibility in mind.

Another case is known where an electric current
certainly consists of two opposite streams of electricity,
viz. the case of the Holtz machine.   While the
machine is being turned, with its terminals somehow
connected, the glass plate acts as a carrier conveying
a charge from one collecting comb to the other at
every half revolution ; but, whereas it carries positive
electricity for one half a rotation, it carries negative

for the other half. The top of the Holtz disk is always, say, positively charged, and is travelling forward, while the bottom half, which is travelling backward at an equal rate, is negatively charged.

In the Holtz case the speeds are necessarily equal, but the charges are not. In the electrolytic case the charges are necessarily equal, but the speeds are not. Each atom has its own rate of motion in a given liquid, independently of what it may happen to have been combined with. This is a law discovered by Kohl-rausch. Hydrogen travels faster than any other kind of atom ; and on the sum of the speeds of the two opposite atoms in a compound the conductivity of the liquid depends. Acids, therefore, in general conduct better than their salts.

37. The following table gives the rates at which atoms of various kinds can make their way through nearly pure water, when urged by a slope of potential of 1 volt per linear centimetre :—

H	.	.	.	1·08	centimetre per hour.
K	.	.	.	0·205	,,
Na	.	.	.	0·126	,,
Li	.	.	.	0·094	,,
Ag	.	.	.	0·166	,,
Cl	.	.	.	0·213	,,
I	.	.	.	0·216	,,
NO_3	.	.	.	0·174	,,

CHAPTER V.

Electrical Inertia.

38. RETURNING now to the general case of conduction, without regard to the special manner of it, we must notice that, if a current of electricity is anything of the nature of a material flow, there would probably be a certain amount of inertia connected with it, so that to start a current with a finite force would take a little time ; and the stoppage of a current would also have either to be gradual or else violent. It is well known that if water is stagnant in a pipe it cannot be quite suddenly set in motion ; and again, if it be in motion, it can only be suddenly stopped by the exercise of very considerable force, which jars and sometimes bursts the pipe. The impetus of running water is utilized in the water-ram. It must naturally occur, therefore, to ask whether any analogous phenomena are experienced with electricity ; and the

answer is, they certainly are. A current does not start instantaneously : it takes a certain time—often very short—to rise to its full strength ; and when started it tends to persist, so that if its circuit be suddenly broken, it refuses to stop quite suddenly, and bursts through the introduced insulating partition with violence and heat. It is this ram or impetus of the electric current which causes the spark seen on breaking a circuit ; and the more sudden the breakage the more violent is the spark apt to be.

The two effects—the delay at making circuit, and the momentum at breaking circuit—used to be called "extra-current" effects, but they are now more commonly spoken of as manifestations of " self-induction."

We shall understand them better directly ; meanwhile they appear to be direct consequences of the inertia of electricity ; and certainly if electricity *were* a fluid possessing inertia it would behave to a superficial observer just in this way.

39. But if an electric current really possessed inertia, as a stream of water does, it would exhibit itself not only by these effects but also mechanically. A conducting coil delicately suspended might experience a rotary kick every time a current was started or stopped in it ; and a coil in which a steady current is maintained should behave like a top or gyrostat, and resist any force tending to deflect its plane.

Clerk Maxwell has carefully looked for this latter form of momentum effect, and found none. He took a bar electro-magnet, mounted it on gimbals so that it was free to rotate if it wished, and then spun it rapidly about an axis perpendicular to the magnetic axis. If there had been the slightest gyrostatic action, the magnet would have rotated about the third perpendicular axis. But it did nothing of the kind. One may say, in fact, that nothing like momentum has yet been observed in an electric current by any *mechanical* mode of examination. A coil or whirl of electricity does not behave in the least like a top (§ 185).

Does this prove that a current has no momentum? By no means necessarily so. It might be taken as suggesting that an electric current consists really of two equal flows in contrary directions, so that mechanically they neutralize one another completely, while electrically—*i.e.* in the phenomena of self-induction or extra-current—they add their effects (§ 89). Or it may mean merely that the momentum is too minute to be so observed. Or, again, the whole thing—the appearance of inertia in some experiments and the absence of it in others—may have to be explained in some altogether less simple manner, to which we will proceed to lead up.

Condition of the Medium near a Circuit.

40. So far we have considered the flow of electricity as a phenomenon occurring solely inside conductors ; just as the flow of water is a phenomenon occurring solely inside pipes. But a number of remarkable facts are known which completely negative this view of the matter. Something is no doubt passing along conductors when a current flows, but the disturbance is not *confined* to the conductor; on the contrary, it spreads more or less through surrounding space.

The facts which prove this have necessarily no hydraulic analogue, but must be treated *suorum generum*, and they are as follows :—

(1) A compass needle anywhere near an electric current is permanently deflected so long as the current lasts.

(2) Two electric currents attract or repel one another, according as they are in the same or opposite directions.

(3) A circuit in which a current is flowing tends to enlarge itself so as to inclose the greatest possible area.

(4) A circuit conveying a current in a magnetic field tends either to enlarge or to shrink or to turn part way round according to the aspect it presents to the field

(5) Conductors in the neighbourhood of an electric circuit experience momentary electric disturbances every time a current in it is started or stopped or varied in strength.

(6) The same thing happens even with a circuit conveying a steady current if the distance between it and a conductor is made to vary.

(7) The effects of self-induction, or extra-currents, can be almost abolished by doubling a covered wire conveying the current closely on itself, or better by laying a direct and return ribbon face to face; whereas they may be intensified by making the circuit inclose a large area, more by coiling it up tightly into close coil, and still more by putting a piece of iron inside the coil so formed.

Nothing like any of these effects is observable with currents of water ; and they prove that the phenomena of the current, so far from being confined to the wire, spread out into space and affect bodies at a considerable distance.

41. Nearly all this class of phenomena were discovered by Ampère and by Faraday, and were called by the latter " current-induction." According to his view, the dielectric medium round a conducting circuit is strained, and subject to stresses, just as is the same medium round an electrically charged body. The one is called an electrostatic strain, the other an electro-magnetic or electro-kinetic strain.

But whereas electrostatic phenomena occur *solely* in the medium—conductors being mere breaks in it, interrupters of its continuity, at whose surface charge-effects occur, but whose substance is completely screened from disturbance—that is not the case with electro-kinetic phenomena. It would be just as erroneous to conceive electro-kinetic phenomena as occurring solely in the insulating medium as it would be to think of them as occurring solely in the conducting wires. The fact is, they occur in both—not only at the surface of the wires, like electrostatic effects, but all through their substance. This is proved by the fact that conductivity increases in simple proportion with sectional area ; it is also proved by every part of a conductor getting hot ; and it is further proved in the case of liquids by their decomposition.

But the equally manifest facts of current attraction and current induction prove that the effect of the current is felt throughout the surrounding medium as well, and that its intensity depends on the nature of that medium ; we are thus wholly prevented from ascribing the phenomenon of self-induction or extra-current to simple and straightforward inertia of electricity in a wire like that of water in a pipe.

We are thus brought face to face with another suggestion to account for these effects, viz. this : Since the molecules of a dielectric are inseparably connected with electricity, and move with it, it is possible that

electricity itself has no inertia at all, but that the
inertia of the atoms of the displaced dielectric confer
upon it the appearance of inertia. Certainly they do
sometimes confer upon it this appearance, as we see
in the oscillatory discharge of a Leyden jar. For a
displaced thing to overshoot its mean position and
oscillate till it has expended all its energy is a proceeding eminently characteristic of inertia ; and so,
perhaps, the phenomena of self-induction are similarly,
though not so simply, explicable (§ 98).

Further consideration of this difficult part of the
subject is, however, best postponed to Part III.
(§§ 48 and 88).

Energy of the Current.

42. I have now called attention to the fact that the
whole region surrounding a circuit is a field of force
in which many of the most important properties of the
current (the magnetic, to wit) manifest themselves.
But directly we begin thus to attend to the whole
space, and not only to the wires and battery, a very
curious question arises. Are we to regard the current
in a conductor as propelled by some sort of end-thrust,
like water or air driven through a pipe by a piston or
a fan, or are we to think of it as propelled by side
forces, a sort of lateral drag, like water driven along a
trough by a blast of air or by the vanes of paddle-

wheels dipping into it? Or, again, referring to the cord models, Figs. 5, 6, and 13, were we right in picturing the driving force of the battery as located and applied where shown in the diagrams, or ought we to have schemed some method for communicating the power of the battery by means of belts or other mechanism to a great number of points of the circuit?

Prof. Poynting has shown that, on the principles developed by Maxwell, the latter of these alternatives, though apparently the more complicated, is the true one ; and he has calculated the actual paths by which the energy is transmitted from the battery to the various points of a circuit, for certain cases.

We must learn, then, to distinguish between the flow of *electricity* and the flow of electric *energy :* they do not occur along the same paths. Hydraulic analogies, at least hydraulic analogies of a simple kind, break down here. When hydraulic power or steam power is conveyed along pipes, the fluid and its energy travel together. Work is done at one end of the tube in forcing in more water, and this is propagated along the tube and reappears at the distant end as the work of the piston. But in electricity it is not so. Electric energy is not to be regarded as pumped in at one end of a conducting wire, and as exuding in equal quantities at the other. The *electricity* does indeed travel thus—whatever the travel of electricity may ultimately be found to mean—but the energy does not. The battery emits its energy, not to the wire direct, but to the surrounding medium ; this is disturbed and strained, and propagates the strain on from point to point till it reaches the wire and is dissipated. This, Prof. Poynting would say, is the function of the wire : it is to dissipate the energy crowding into it from the medium, which else would take up a static state of strain and cease to transmit any more. It is by the continuous dissipation of the medium's energy into heat that continuous propagation is rendered possible (§ 107).

The energy of a dynamo does not therefore travel to a distant motor through the wires, but through the air. The energy of an Atlantic cable battery does not travel to America through the wire strands, but through the insulating sheath. This is a singular and apparently paradoxical view, yet it is well founded.

Think of a tram-car drawn by an underground rope, like those in the streets of Chicago or Hampstead Hill. A contact piece of iron protrudes from the bottom of the car and grips the moving rope, which is thus enabled to propel the car. How does the energy of the distant stationary engine reach the car ? *Viâ* the rope and the iron connector, undoubtedly. They both have to be strong, and are liable to be broken by the transmitted stress.

Next, think of an electric tram-car driven by means

of a current taken up from an underground conductor, like that of Mr. Holroyd Smith at Manchester, or at the late Inventions Exhibition. A contact piece of wire rope protrudes from the bottom of the car and drags a little truck along the conductor, which is thus enabled to supply electricity to the electro-magnetic motor geared to the wheels. How does the energy of the distant dynamo reach the car in this case ? *Not* *viâ* the wire connector ; not even *viâ* the underground conductor. It travels from the distant dynamo through the general insulating medium between cable and earth, some little enters the conductor and is dissipated, but the great bulk flows on and converges upon the motor in the car, which is thus propelled. All the energy of the conducting wire is dissipated and lost as heat : it is the energy of the insulating medium which is really transmitted and utilized.

The manner in which the transmission of energy goes on we will attend to further in Part III. (§ 105 *et seq.*).

Phenomena Peculiar to a Starting, or Stopping, or Varying Current.

43. There is a remarkable fact concerning electric currents of varying strength, which has been lately brought into prominence by the experimental skill of Prof. Hughes, viz. that a current does not start or

H

stop equally and simultaneously at all points in the
section of a conductor, but starts at the outside first.
This fact is naturally more noticeable with thick wires
than with thin, and it is especially marked in *iron* wires,
for reasons which in Part III. will become apparent;
but the general cause of it in ordinary copper wires
can very easily be perceived in the light of the views
of Prof. Poynting just mentioned.

For, remember that a current in a wire is not
pushed along by a force applied at its end, so as to be
driven over obstacles by its own momentum combined
with a *vis a tergo ;* but it is urged along at every point
of its course by a force just sufficient to make it overcome the resistance there, and no more, the force being
applied to it through the medium of the dielectric
in which the wire is immersed. A lateral force it is
which propels the electricity ; and it naturally acts
first on the outer layers of the wire or rod, only acting
on the interior portions through the medium of the
outside (§ 102).

44. To illustrate this matter further, rotate a
common tumbler of liquid steadily for some time and
watch the liquid ; dusting powder perhaps over it to
make it more visible. You will see first the outer
layer begin to participate in the motion, and then the
next, and then the next, and so on, until at length the
whole is in rotation. Stop the tumbler, and the liquid
also begins gradually to stop by a converse process :

the outside stopping first, and then gradually the central portions.

If the liquid sticks together pretty well, like treacle, the motion spreads very rapidly: this corresponds to a poor conductor. If the liquid be very mobile, the propagation of motion inward is slow: this corresponds to a very good conductor. If the liquid were perfectly non-viscous, it would correspond to a perfect conductor, and no motion would ever be communicated to it deeper than its extreme outer skin.

Think now of an endless tube full of water, say the hollow circumference of a wheel, or the rim of a top, and spin it: the liquid is soon set in rotation, especially if the tube be narrow or the liquid viscous; but it is set in motion by a lateral not an end force, and its outer layers start first.

Just so is it with a current starting in a metal wire. If the wire be fine, or its substance badly conducting, it all starts nearly together; but if it be made pretty thick, and of well conducting substance, its outer layers may start appreciably sooner than the interior. And if it were infinitely conducting, no more than the outer skin would ever start at all (see Chap. X. and § 103).

In actual practice the time taken for all the electricity in an ordinary wire to get into motion is excessively short—something like the thousandth of a second—so that the only way to notice the effect is to start and reverse the current many times in succession.

45. If the hollow-rimmed wheel above spoken of were made to oscillate rapidly, it is easy to see that only the outer layers of water in it would be moved to and fro ; the innermost water would remain stationary ; and accordingly it would appear as if the tube contained much less water than it really does. The virtual bore of the pipe would, in fact, for many purposes be diminished.   So is it also with electricity ; the sectional area of a wire to a rapidly alternating current is virtually lessened so far as its conducting power is concerned ; and accordingly its apparent resistance is higher for alternating than for steady currents. The effect is, however, too small to notice in practice, except with thick wires and very rapid alternations.

By splitting up the conductor into a bundle of insulated wires, thus affording the dielectric access to a considerable surface of conductor, the force is applied much more thoroughly, and so the effect spoken of is greatly lessened.   The same thing is achieved by rolling out the conducting-rod into a flat thin bar. Making the conductor hollow instead of solid offers no particular advantage, beyond the gain of surface per given weight, because no energy travels *viâ* the hollow space, it still. arrives only from the outside ; *unless*, indeed, the return part of the circuit is taken along the axis of the hollow like a telegraph cable.

In this last arrangement all the energy travels *viâ* the dielectric between the two conductors, and none travels outside at all. It will be perceived therefore that, as in static electricity, the term "outside" must be used with circumspection : it really means that side of a conductor which faces the opposite conductor across a certain thickness of dielectric.

46. We learn from this that, whereas in the case of steady currents the sectional area and material of a conductor are all that need be attended to, the case is different when one has to deal with rapidly alternating currents, such as occur in a telephone, or, again, such as are apt to occur in a Leyden-jar discharge (see Part I., p. 41), or in lightning.

In all these cases it is well to make the conductor expose considerable surface to the propelling medium —the dielectric—else will great portions of it be useless.

Hence, a lightning-conductor should not be a round rod, but a flat strip, or a strand of wires, with the strands as well separated as convenient.

47. I might go on to say here that iron makes an enormously worse conductor than copper for rapidly alternating currents. So it does for currents which alternate with moderate rapidity—a few hundred or thousand a second—like those from a dynamo or a telephone ; but, singularly enough, when the rapidity of oscillation is immensely high, as it is in Leyden-jar

discharges and lightning, iron is every bit as good as copper, because the current keeps to the extreme outer layer of the conductor, and so practically does not find out what it is made of.

The Question of Electrical Momentum again.

48. We are now able to return to the important question whether an electric current has any momentum or not, as it would have if it were a flow of

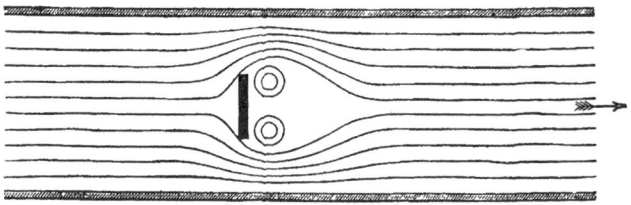

FIG. 14.—Stream-lines of water flowing through a pipe with an obstruction in it.

material liquid. Referring to Part I. § 7, a hint will be found that the laws of flow of a current in conductors—the shape of the stream-lines, in fact—are such as indicate no inertia, or else no friction. Now Ohm's law shows that at any rate *friction* is not absent from a current flowing through a metal ; hence it would appear at first sight as if *inertia* must be absent.

The stream-lines bear upon the question in the

following kind of way.   If an obstacle is interposed
in the path of a current of water, the motion of the
water is unsymmetrical before and behind the obstacle.
The stream-lines spread out as the water reaches the
obstacle, and then curl round it, leaving a space full
of eddies in its wake (Fig. 14).

But if one puts an obstacle in the path of an electric
current—say by cutting a slit in a conducting strip of
tinfoil—the stream-lines on either side of it are quite
symmetrical, thus—

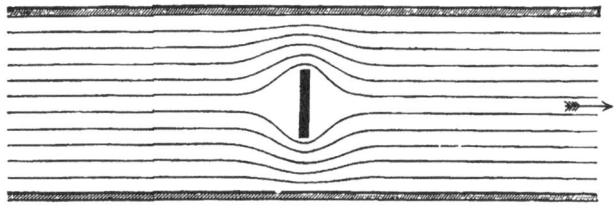

FIG. 15.—Electrical stream-lines past an obstacle.

And this is exactly what would be true for water
also, if only it were devoid either of friction or of
inertia, or of both.

49. Is not this fact conclusive, then ?   Does it not
prove the absence of momentum in electricity ?

Plainly the answer must depend on whether there is
any other possible mode of accounting for this kind of
flow.   And there is.

For suppose that water, instead of being urged by

something not located at or near the obstacle—instead
of being left to its own impetus to curl round or shoot
past as it pleases—suppose it were propelled by a force
acting at every point of its journey, a force just able
to drive it at any point against the friction existing
at that point and no more ; then the flow of water
would take place according to the electrical stream-
lines shown in Fig. 15.

An illustration of such a case is ready to hand.
Take a spade-shaped piece of copper wire or sheet,
heat it a little, and fix it in quiescent smoky air ;
looking along it through a magnifier in a strong light
you will see the warmed air streaming up past the
metal according to the stream-lines of Fig. 15 ; and
this just because the moving force has its location at
the metal surface, and not in some region below it.[1]
One cannot indeed say that it is propelled at every
point of its course, but it is propelled at the critical
points where the special friction occurs, and this
comes to sufficiently the same thing.

We learn, therefore, that stream-lines like Fig. 15
prove one of three things, not one of two ; and the
three things are : (1) that the fluid has no friction ;
or (2) that it has no inertia ; or (3) that it is propelled
at every point of its course.

If any one of these is true of electricity, there is no
need to assume either of the others in order to explain

[1] See Lord Rayleigh, *Nature*, vol. xxviii. p. 139.

the actual manner of its flow. Now we have just seen in § 42 that, according to Prof. Poynting's interpretation of Maxwell's theory, the third of the above is true—electricity is propelled at every point of its course ; consequently, as said in Part I..§ 7, the question of its inertia so far remains completely open (§§ 88, 89, and 98).

CHAPTER VI.

CHEMICAL AND THERMAL METHODS OF PRODUCING CURRENTS. CONDUCTION IN GASES.

Voltaic Battery.

50. Leaving this singular mode of regarding the subject for the present, to return to it in Chap. X. Part III., let us proceed to ask how it comes about that a common battery or a thermopile is able to produce a current (read Chapter IV. again).

If we allow ourselves to assume the existence of an unexplained chemical attraction between the atoms of different substances, an explanation of the action of an ordinary battery cell is easy. You have first the liquid containing, let us say, hydrogen and oxygen atoms, free or potentially free—that is, either actually dissociated, or so frequently interchanging at random from molecule to molecule that the direction of their motion may be guided by a feeble directive force (§ 33). Each of these atoms in the free state possesses a charge of electricity—the hydrogen all a certain amount of positive electricity, the oxygen twice that amount of negative. Into this liquid you then plunge a couple of metals which attract these atoms differently: for instance, zinc and copper, which both attract oxygen, but zinc more than copper; or, better, zinc and platinum, the latter of which hardly attracts it at all ; or, better still, zinc and peroxide of lead, one of which attracts oxygen, the other hydrogen.

Immediately, the free oxygen atoms begin moving up to the zinc, the free hydrogen atoms to the other plate.

51. When one speaks of the plates attracting the atoms, it is not necessary to think of their exerting a force on all those in the liquid, distant and near : all that is necessary is to assume a force acting on those which come within what is called " molecular range " of its surface—a distance extremely minute, and believed to be about the ten-millionth part of a millimetre. If the zinc plate removes and combines with all the oxygen atoms which come within this range, they will be speedily replaced by others from the next more distant layer by diffusion, and these again by others, and so on. And thus there will be a gradual procession of oxygen atoms all through the liquid towards the zinc, the rate of the procession being regulated by the force acting, and by the rate of diffusion possible in the particular liquid used. All the atoms which reach the zinc neutralize a certain

portion of its electricity by means of the positive charge they carry, and thus very soon it would become positively electrified enough to neutralize its attractive power on the similarly charged oxygen atoms, and everything would stop. But if a channel for the escape of its electricity be provided by leading a wire from it to the copper plate, the circuit is completed, the electricity streams back by the wire, and the procession goes steadily on. The electricity thus imparted to the copper, or platinum, neutralizes any repulsion it exerted on the negatively charged hydrogen atoms, and make them in a similar way begin a procession towards it, deliver up their charges to it, combine with each other, and escape as gas.

Without going into all the niceties possible, this mode of thinking of the matter at least calls attention to some of the more salient features of a battery.

52. If, instead of two different plates, plates of the *same* metal be immersed, they will need to be oppositely electrified by some means before they are able to cause the two opposite processions, and so maintain a current in the liquid. This plainly corresponds to a voltameter.

53. Taking advantage of the known fact that the atoms are charged, Helmholtz avoids the necessity for postulating any chemical (non-electrical) force between zinc and oxygen, by imagining that all substances have a specific attraction for electricity

itself, and that zinc exceeds copper and the other common metals in this respect.

He would thus think of the zinc attracting, not the oxygen itself, but its electric charge ; and so would liken a battery cell still more completely to a voltameter. The polarization or opposition force acting at the hydrogen-evolving plate he would account for by the attraction of hydrogen for negative electricity, and the consequent repugnance of the hydrogen atoms to part with their charges.

Volta's so-called Contact Force.

54. It may be convenient to append to this account of the action of a battery a statement of the way in which the electric charges observed on plates of zinc and copper which have been put into contact and separated are brought about. It is a very simple matter, though a great deal has been written about it.

Plates of zinc and copper immersed in air are under precisely the same chemical conditions as if they were immersed in water. The only difference is that, whereas water is a conductor, air is an insulator. Until the plates of zinc and copper (or other pair of metals) are made to touch, nothing happens in either case, because the chemical tendency is uniform all over both plates ; and though the attraction of the

zinc for oxygen is pretty strong it would be impossible
for it to combine with many atoms, receiving their
charges, without becoming so negatively charged as
to repel them electrically as much as it attracts them
chemically. This, indeed, may be considered as the
state of equilibrium, which is instantaneously attained.

But directly metallic contact between the two
metals is effected, all the oxygen atoms at this point
are swept away, and a clear passage is opened from
the zinc to the copper for the flow of electricity.
Unless, therefore, there is some E.M.F. at their junction
—which we have good reason for asserting there is *not*,
of any magnitude worth speaking of—an immediate
rush of negative electricity from zinc to copper, or of
positive the other way, occurs. The copper therefore
becomes negatively charged, the zinc becomes positive.
So far everything goes on just the same whether the
plates are in acidulated water or in common air.

What happens next depends upon the difference
between water and air in conducting power—that is,
in the existence of the potentially free or dissociated
atoms necessary to electrolytic conduction. Acidulated
water possesses these; air does not. Accordingly,
in the case of the water the negative charge is being
continually carried back from the copper to the zinc
by a procession of oxygen atoms, which continually
are drawn up against the zinc and combine with it,
as already explained ; whereas in the air nothing

further happens except the slight electrostatic strain into which the air is thrown by the quantity of electricity accumulated upon the metals, positive on the zinc, negative on the copper, and which has no vent or outlet. Ordinarily these charges will be extremely small, the electromotive force producing them being rather under than over 1 volt, and accordingly the electrostatic strain in the air near a couple of zinc and copper plates in contact is extremely minute. By suspending a delicate aluminium needle highly charged near such a junction, however, Sir William Thomson has been able to observe the state of strain : the needle if positively charged moving perceptibly towards the copper. The more usual method of rendering the phenomenon conspicuous, and the one originally used by Volta, is to increase the capacity of the arrangement by bringing two carefully ground plates very close together. Although the E.M.F. is small (just the same as with a mere point contact), yet now the capacity is so great that quite a reasonable quantity of electricity can be stored in the two opposing metals, opposing each other across an air film and only really touching at a few points ; so that when they are neatly separated sufficient charge is found in them to affect even a common gold-leaf electroscope.

55. The mistake which has been, and still frequently is, made with regard to this simple and not very important experiment, has been to regard the charge

as evidence of a peculiar E.M.F., at the point of contact, causing a difference of potential in the two metals. And this fictitious contact E.M.F. has then been appealed to to explain the voltaic battery.

The right way of regarding the matter is to consider the battery first, and explain its action chemically so far as it is possible to explain it at present; and then to point out that similar things will occur in air (an air battery, in fact), with the slight difference that since air is a dielectric instead of an electrolyte no continuous current is possible, but merely a slight electric displacement.

56. The effective cause of the whole phenomenon in either case is the greater affinity of oxygen for zinc rather than copper. This by itself would cause a greater strain of negative electricity towards zinc—a slackening of the negative cords in it, to speak in the language of the cord model—and a consequent rise of negative potential. A piece of isolated zinc is therefore some 1·8 volt below the potential of the atmosphere. The same sort of thing is true for copper except that the intensity of strain is less; as evidenced by the less heat of formation of CuO compared with ZnO; and accordingly a piece of isolated copper is about 0·8 volt below the potential of the atmosphere.

Directly the two metals touch they necessarily become of the same potential—all parts of a conductor are at one potential unless there are disturbing internal forces—and the equalization of potential is effected by the rush of electricity across the junction, whereby the zinc receives a positive charge and the copper a negative charge, until their potential is equalized. In air the equalization is effected in an instant. In water it is a matter of eternity. That is all the difference. The thing observed in the Volta effect is not a difference of *potential* between zinc and copper, but a

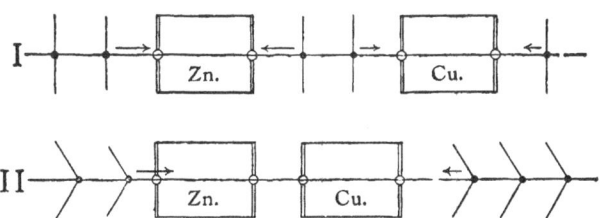

FIG. 15A.—Diagrammatic representation of the Volta effect on the plan of the cord models (Figs. 5, 6, 7, &c.).

I. shows a piece of zinc and copper before contact, with a cord representing negative electricity passing through both, and beads representing oxygen atoms. The arrows indicate that the oxygen is being pulled by the zinc on all sides of it; and that it is also pulled by the copper but with less force.

II. shows the effect of sweeping away the oxygen atoms between the two metals and establishing metallic contact, so that the greater atom-attracting force of zinc over copper can now produce an effect until it is balanced by the elastic stress called out by an electric displacement. The surface of the zinc has now less than its normal share of negative cord—it is positively charged. The copper is negatively charged. This is the Volta effect.

difference of *charge ;* the two metals being charged so as to make their potential uniform.

What is observed in the Sir William Thomson form of the experiment is again not a difference of potential between zinc and copper, but a slope of

I

potential in the air near them, from the zinc towards the copper. The metals when in contact are both at a common potential, 1·3 volts below the atmosphere, the mean of their original potentials, but the original difference of potential between each and the air in contact with it remains unaltered; hence there is a gradual slope of potential of 1 volt from the layer of air in contact with zinc to the layer in contact with copper; and this slope of potential is what the electrometer needle feels. The diagram Fig. 15A may possibly help in making the thing clear.

True Contact Force.

57. So far we have assumed that there is actually *no* force at the contact of zinc with copper. There is indeed none of any appreciable magnitude, but the force is not absolutely zero. Between some metals, bismuth and antimony for example, the force is much larger, but it is still only a few hundredths of a volt. It is an important thing that there can be a true contact force at the junction of two metals, only it has nothing to do with the chemically produced Volta effect. If the Volta effect be called a contact force at all, it is a contact force between metal and air; any true contact force between metals acts as a slight and insignificant disturber of

the simple Volta effect, and what is really observed in electroscopic experiments is the sum of the two.

58. That there is a true though weak contact force at the junction of metals is proved by the reversible heat effects which are found there when a current is passed across the junction : a current one way produces more heat than a current the other way. In a simple homogeneous piece of metal the heat produced by a current is utterly independent of direction : it is called irreversible heat ; it is proportional to the square of the current strength, as Joule showed. But at a junction of different substances, or even at a junction of the same substance in two different states—two different temperatures, for example,—in addition to the irreversible heat produced by mere resistance there is a reversible heat production, one which changes sign with the direction of the current, so that the current one way actually tends to cool the junction instead of heating it. With care this may be got to overpower and mask the irreversible heat, and a junction may be cooled and water frozen by steadily passing a moderate current in the right direction across it. This curious effect was discovered by Peltier.

It may be considered as the fundamental fact of thermo-electricity. Its meaning is that something in the metals at the junction is helping to propel the current along ; doing work in fact, and consuming its

own heat in the process. The vibratory motion
of the molecules is getting used up in propelling electricity. The contact force is acting in the direction
of the current.

If the current be reversed, it will be driven
against the force of the molecules, and an extra
amount of heat will be added to the irreversible or
frictional generation of heat.

59. This thermal evidence of contact force, though
the most direct, was not the earliest discovered. The
earliest known fact in thermo-electricity was that in a
complete circuit of different metals a current could be
excited by having the parts at different temperatures;
manifestly because these contact forces we have
been speaking of change with temperature—some
increasing, others decreasing. They are accurately
balanced in a circuit of uniform temperature, but
they have a resultant whenever the temperature is
not uniform, and this resultant propels the current
discovered by Seebeck.

Thermo-electric Pile.

60. A thermopile may be thought of in the
following way, but in trying to understand the nature
of these actions at present one must admit that some
speculation and vagueness exist.

We have seen that when electricity is propelled

through or among the molecules of a metal it experiences a certain resistance or opposition force which is exactly proportional to the speed of its motion (§ 28). In other words, there is a connection between matter and electricity in many respects analogous to fluid friction but varying accurately as the first power of the relative velocity. Hence, if an atom of matter be vibrating about a fixed point, it will tend to drive electricity to and fro with it; but if it be only one of a multitude, all quivering in different phases, they will none of them achieve any propulsion. This may be considered the state of an ordinary warm solid. But if from any cause a set of atoms could be made to move faster in one direction than in the reverse direction—to move forwards quickly and backwards slowly—then such an unsymmetrically-moving set *will* exert a propulsive tendency and tend to drive a current of electricity forwards, simply because the force exerted is proportional to the velocity, and so is greater on the forward journey than on the return. Referring back to the cord model, Fig. 5, Ohm's law requires that the friction between cord and beads should be directly as the velocity, hence if a bead begins to oscillate unsymmetrically, travelling forward quickly and back slowly, it will propel the cord along in the direction in which it moves most quickly, somewhat as a child can propel its chair by jogging it upon a rough floor.

Wherever conduction of heat is going on along a substance the atoms are in this condition. They are driven forward infinitesimally quicker by the more rapidly moving atoms at the hot end, than they are driven back by the less rapidly moving atoms in front. And hence such a slope of temperature exerts a propulsive tendency: there is an electromotive force in a substance unequally heated.

This fact was discovered theoretically and verified experimentally by Sir William Thomson.

61. But not only is there such a force at a junction of a hot and cold substance ; there is also a force at the junction of two substances of different kinds, even though the temperature be uniform. It is not quite so easy to explain how it now comes about that the atoms at this kind of junction are moving faster one way than the other ; nevertheless, such a thing is not unlikely, considering the state of constraint and accommodation which must necessarily exist at the boundary surface of two different media. However it be caused, there is certainly an E.M.F. at such a junction.

Thus, then, in a simple circuit of two metals, with their junctions at different temperatures, there are altogether four electromotive forces—one in each metal, from hot to cold or *vice versâ*, and one at each junction ; and the current which flows round such a circuit is propelled by the resultant of these four.

These four forces, two Thomson forces in the metals, and two Peltier forces at their junctions, may some of them help and some hinder the current. Whenever they help, the locality is to that extent cooled; whenever they hinder, it is to that extent warmed.

Frictional Electricity.

62. But the contact force at a junction is by no means confined to metals. It occurs between insulators also, and it is to it that the striking effects produced by all frictional electric machines are due. The essential thing in the production of "frictional electricity" is the contact of dissimilar substances. It is by their contact force that electricity gets transferred from one to the other so that one becomes positive and the other negative. The violence of friction is sometimes necessary to aid the transfer; the substances being so badly conducting.

By thus noticing that the connection between matter and electricity, known as resistance and defined by Ohm's law (§ 28), is competent to produce contact electromotive forces, we may perceive how it comes to pass that in good conductors such forces are so weak, while in insulators they are so strong. Electricity slips through the fingers of a metal as it were, and the driving force it can exert is very feeble; while an insulator gets a good grip and thrusts it along with violence.

*Specific Relation between Matter and Electricity ;
sometimes called " Specific Heat of Electricity."*

63. The metals differ in their gripping power,
and, roughly speaking, the best conductor makes
the worst thermo-electric substance. A bad conductor, like antimony, or, still better, galena, or selenium,
or tellurium, makes a far more effective thermo-
electric element than a well-conducting metal. Not
that specific resistance is all that has to be considered in the matter ; there is also a specific relation
between each metal and the two kinds of electricity.
Thus, iron is a metal whose atoms have a better
grip of positive than of negative electricity, and so
a positive current gets propelled in iron from hot to
cold. Copper, on the other hand, acts similarly on
negative electricity, and it is a negative current
which is driven from hot to cold in copper. And all
the metals can be classed with one or other of these
two, except perhaps lead, which appears to grip
both equally, and so to exert no differential effect
upon either.

How this relation can be likened to a "specific
heat," may be thought out by attending to the last
paragraph of § 61, and by regarding electricity as a
material fluid (see also § 182).

Pyro-electricity.

Certain crystals, called by mineralogists hemihedral, having different forms at the two ends of their axis, which may be called the A end and the B end respectively, exhibit some properties not quite the same in the direction A B as in the direction B A. They are more easily scratched, for instance, in one sense than in the other. Such crystals, of which the class of tourmalines may be taken as the type, have other very singular properties. Such of them as are fairly clear are opaque to light in a singular fashion—not opaque to light polarized in all planes, but selectively opaque. Vibrations occurring perpendicular to the axis are rapidly quenched, so that one cannot see at all through a slice taken perpendicular to the axis, while vibrations occurring along the axis are transmitted with but moderate absorption. This opacity seems quite different from the conducting opacity of metals, about which we shall speak later, for, in the first place, the light stopped is not reflected, but absorbed ; and, in the second place, a crystal of tourmaline is not a conductor, but a very fair insulator.

And yet there are some peculiarities about such conducting power as it has which are very note-

worthy, and which may be intimately connected with
the selective opacity which fits a slice of crystal cut
parallel to the axis for use as a "polarizer" in optics.
One of these peculiarities was found by Dr. S. P.
Thompson in conjunction with the present writer, viz.
that while, like all other uniaxial crystals, the con-
ductivities for heat along and across the axis are not
the same (being, in the case of tourmaline, less good
along the axis than across), yet, in addition to this,
a warming crystal conducts heat better in the sense
B A than in the sense A B, while a cooling crystal
does the opposite. While the temperature is rising
their heat gets conveyed more easily towards A than
towards B.

Whether on account of inequalities of temperature
thus set up, or for some more direct reason, the same
is true of electricity. And accordingly, while a crystal
is rising in temperature, positive electricity accumu-
lates at the A end, and negative electricity at the B
end. So long as the temperature remains constant
nothing further happens, except ordinary leakage,
principally no doubt over the surface, which may in
time completely mask the effect produced. On now
cooling the crystal, an inverse electrification will
be set up; or, if no leakage had been permitted,
the effect of cooling will be to simply replace the
electricity displaced by the warming.

While the crystal is steady in temperature no perceptible difference in electric conductivity has been detected by the writer between the sense A B and the sense B A. Neither is there any difference in the thermal conductivity when the temperature is steady. Both effects depend on a varying temperature.

Passage of Electricity through a Gas.

64. There remains to be said something about the way in which electricity can be conveyed by *gases.*

The first thing to notice is that there is no true conduction through either gases or vapours ; in other words, a substance in this condition seems to behave as a perfect insulator—perhaps the only perfect insulator there is. Not water vapour, not even mercury vapour, is found to conduct in the least. This shows that mere bombardment of molecules, such as is known to go on in gases, is not sufficient either to remove or to impart any electric charge.

The commonest way in which electricity makes its way through a gas, setting aside the mere mechanical conveyance by solid carrier, is that of disruptive discharge. Let us try and look into the manner of this a little more closely, if possible.

First of all, since locomotion is possible to the molecules of a gas the same as of any other fluid, it is natural to ask why electrolysis does not go on as in a liquid. Now, for electrolysis in a liquid two conditions seemed necessary : first, that the atoms or radicles in a molecule should be oppositely charged with electricity ; second, that they should be in such a condition (whether by dissociation or otherwise) that interchanges of atoms from molecule to molecule, or, in some other way, a procession of atoms, could be directed in a given direction by a very feeble or infinitesimal force.

Since a gas does *not* act as an electrolyte, one of these conditions, or perhaps both, must fail. Either the atoms of a gas-molecule are not charged, which is a plausible hypothesis for elementary gases, or else the atoms belonging to a gas-molecule remain individually belonging to it, and are not readily passed on from one to another.

When one says that a gas does not act as a common electrolyte, the experimental grounds of the statement are that a finite electrostatic stress certainly is possible in its interior—a stress of very considerable amount ; and when this stress does overstep the mark and cause the material to yield, the yielding is evidently not a quiet and steady glide or procession, but a violent breaking down and collapse, due to insufficient tenacity of something. One may therefore picture the molecules of a gas, between two opposite electrodes or discharge terminals maintained at some great difference of potential, as arranged in a set of parallel chains from one to the other, and strained nearly up to the verge of being torn asunder. In making this picture one need not suppose any fixture of individual molecules : there may be a strong wind blowing between the plates ; but all molecules as they come into the field must experience the stress, and be relieved as they pass out.

65. If the applied slope of potential overstep a certain limit, fixed by observation at something like 33,000 volts per linear centimetre for common air, the molecules give way, the atoms with their charges rush across to the plates, and discharge has occurred. The number of atoms thus torn free and made able to convey a charge by locomotion is so great that there has never been found any difficulty in conveying any amount of electricity by their means. In other words, *during* discharge the gas becomes a conductor, and, being a conductor by reason of locomotion of atoms, it may be called an electrolytic conductor.

But whether the charge then possessed by each carrier atom intrinsically belonged to it all the time, or whether it was conferred upon the components of the molecules during the strain and the disruption, is a point not yet decided.

What is called "the dielectric strength" of a gas— that is, the strain it can bear without suffering disruption and becoming for the instant a conductor—

depends partly on the nature of the gas, and very largely on its pressure. Roughly, one may say that a gas at high pressure is very strong, a gas at low pressure very weak. An ordinary electrolyte might be called a dielectric of zero strength.

One reason why pressure affects the dielectric tenacity of a gas readily occurs to one : it is certainly not the only one, but it can hardly help being at least partially a *vera causa ;* and that is, the fact that in a rare gas there are fewer molecules between the plates to share the strain between them.

Thus if 40,000 volts per centimetre break down ordinary air, 40 volts per centimetre ought to be enough to effect discharge through air at a pressure of about $\frac{3}{4}$ millimetre of mercury ; and at a pressure of 50 atmospheres 2,000,000 volts per centimetre should be needed.[1]

A Current regarded as a Moving Charge.

66. To review the ground we have covered so far. We first tried to get some conception of the nature of electrostatic charge, and the function of a dielectric medium in static electricity. We next proceeded to

---

[1] It is true that tension per unit area, or energy per unit volume, is proportional to the *square* of the potential-slope, and I attach no special importance to the simple proportion assumed in the text. There is a great deal more to be said on these subjects, but this is scarcely the proper lace to say it.

see how far the phenomena of current electricity could
be explained by reference to electrostatics. For a
current, being merely electricity in locomotion, need
consist of nothing but a charged body borne rapidly
along.

Charge a sphere with either positive or negative
electricity, and throw it in some direction ; this constitutes a positive or a negative current in that
direction. There is nothing necessarily more occult
than that. And a continuous current between two
bodies may be kept up by having a lot of pith
balls, or dust particles, oscillating from one to the
other, and so carrying positive electricity one way,
and negative the other way. But such carriers, as
they pass each other with their opposite charges,
would be very apt to cling together and combine.
They might be torn asunder again electrically, or
they might be knocked asunder by collision with
others. Unless they were one or other, the current
would shortly have to cease, and nothing but a
polarized medium would result.

Instead of pith balls, picture charged atoms as so
acting, and we have a rough image of what is going
on in an electrolyte on the one hand, and a dielectric on the other. The behaviour of metals and
solid conductors is more obscure. Locomotive carriage is not to be thought of in them ; but, inasmuch as no new phenomenon appears in their case,

it is natural to try and picture the process as one not wholly dissimilar ; and this is what in § 27 we tried to do ; with, however, but poor success.

67. I have said that an electric current need be nothing more occult than is a charged sphere moving rapidly ; and a good deal has been made out concerning currents by minutely discussing all that happens in such a case. But, even so, the problem is far from being a simple one. One has to consider not only the obviously moving charge, but also the opposite induced charge tied to it by lines of force (or tubes of induction, as they are sometimes called), and we have this whole complicated system in motion. And the effect of this motion is to set up a new phenomenon in the medium altogether—a spinning kind of motion that would not naturally have been expected ; whereby two similarly charged spheres in motion repel one another less than when stationary, and may even begin to attract, if moving fast enough ; whereby also a relation arises between electricity and magnetism, and the moving charged body deflects a compass needle (§§ 113 and 184). Of which more in the next Part.

PART III.

MAGNETISM.

CHAPTER VII.

RELATION OF MAGNETISM TO ELECTRICITY.

68. WE next proceed to consider electricity in a state of *rotation*. What happens if we make a whirl-pool of electricity? Coil up a wire conveying a current, and try. The result is it behaves like a magnet: compass-needles near it are affected, steel put near it gets magnetized, and iron nails or filings get attracted by it—sucked up into it if the current be strong enough. In short, it *is* a magnet. Not of course a permanent one, but a temporary one, lasting as long as the current flows. It is thus suggested that magnetism may perhaps be simply electricity in rotation. Let us work out this idea more fully.

First of all, one may notice that everything that can be done with a permanent magnet can be imitated by a coiled wire conveying a current. (It would not do altogether to make the converse statement.) Float a coil attached to a battery vertically on water, and

K 2

you have a compass-needle : it sets itself with its axis
north and south. Suspend two coils, and they will
attract or repel or turn each other round just like two
magnets.

69. As long as one only considers the action of a
coil at some distance from itself, there is no need to
trouble about the shape of the particular magnet
which it most closely simulates ; but as soon as one
begins to consider the action of a coil on things close

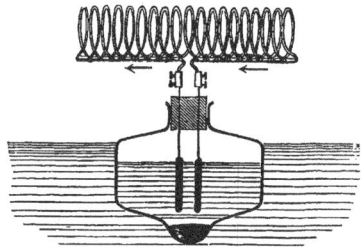

FIG. 16.—Floating battery and helix acting as a compass-needle.

to it, it is necessary to specify the shape of the
corresponding magnet.

If the coil be a long cylindrical helix like a close-
spired corkscrew, as in Fig. 16, it behaves like a
cylindrical magnet filling the same space. If the coil
be a short wide hank, like a curtain-ring, it behaves
again like a cylindrical magnet, but one so short that
it is more easily thought of as a disk. A disk or
plate of steel magnetized with one face all north and

the other face all south can be cut to imitate any thin hank of wire conveying a current. It will be round if the coil be round, square if it be square, and irregular in outline if the coil be irregular.

There is no need for the coil to have a great number of turns of wire except to increase its power : one is sufficient, and it may be of any shape or size. So when we come to remember that every current of electricity must necessarily flow in a closed circuit, one perceives that *every current of electricity is virtually a coil of more or less fantastic shape*, and accordingly imitates some magnet or other which can be specified. Thus we learn that every current of electricity must exhibit magnetic phenomena : the two are inseparable—a very important truth. See Appendix (*a*).

There is one detail in which the magnetized disk and the coil are not equivalent, and the advantage lies on the side of the coil : it has a property beyond that possessed by any ordinary magnet. It has a penetrable interior, which the magnet has not. For space outside both, they simulate each other exactly ; for space inside either, they behave differently. The coil can be made to do all that the magnet can do ; but the magnet cannot in every respect imitate and replace the coil : else would perpetual motion be an every-day occurrence.

70. Now I want to illustrate and bring home

forcibly the fact that there is something rotatory about magnetism—something in its nature which makes rotation an easy and natural effect to obtain if one goes about it properly.   One will not observe this by taking two magnets : one will see it better by taking a current and a magnet, and studying their mutual action.

A magnet involves, as you know, two poles—a north and a south pole—of precisely opposite properties : it may be considered as composed of these two poles for many purposes ; and the action of a current on a magnet may be discussed as compounded of its action on each pole separately.   Now how does a current act on a magnetic pole ?   Two currents attract or repel each other ; two poles attract or repel each other ; but a current and a pole exert a mutual force which is neither attraction nor repulsion : it is a rotatory force.   They tend neither to approach nor to recede ; they tend to revolve round each other.   A singular action this, and at first sight unique.   All ordinary actions and reactions between two bodies take place in the line joining them : the force between a current and a pole acts exactly at right angles to the line joining them.

Helmholtz long ago (in 1847) showed that the conservation of energy could only be true if forces between bodies varied in some way with distance and acted in the line joining them.   Now here is a case where the force is not in the line joining the bodies, and accordingly the conservation of energy is defied : the two things will revolve round each other for ever. This affords, and has afforded, a fine field for the perpetual motionist ; and if only the current would maintain itself without a sustaining power, perpetual motion would in fact be attained. But this after all is scarcely remarkable, for the same may be said of a sewing-machine or any other piece of mechanism : if only it would continue to go without sustaining power it would be a perpetual motion. Attend to pole and current only, and energy is *not* conserved, it is perpetually being wasted ; but include the battery as an essential part of the complete system, and the mystery disappears : everything is perfectly regular.

71. The easiest way perhaps of showing the rotation of a conductor conveying a current round a magnetic pole is to take an 8-feet-long piece of gold thread, such as is stitched upon the garments of military officers, and hanging it vertically supply it with as strong a current as it will stand. Then bring near it a vertical bar-magnet, and instantly you will see the thread coil itself into a spiral, half of it twisting round the north end of the bar, and half twisting as part of the same spiral round the south end (Fig. 17).

If the magnet were flexible and the conductor rigid, instead of *vice versâ*, the magnet would in like manner coil itself in a spiral round the current : the force is

strictly mutual. A rigid magnet, put near a stiff conductor, shows only the last remnants of this action :
it sets itself at right angles to the wire, and ap-
proaches its middle to touch it, but that is all it
can do.

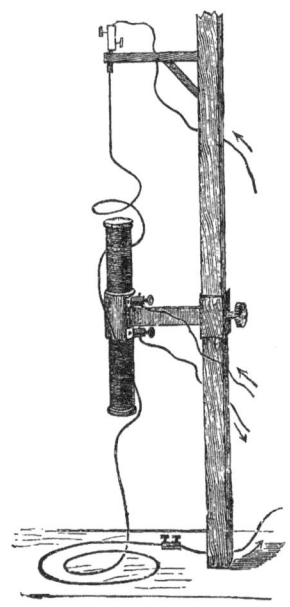

FIG 17.—A long flexible conductor twisting itself into a spiral round a powerful
bar-magnet raised to meet it.

The experiment with the flexible gold thread is
simple, satisfactory, and striking, but the rotatory
properties connected with a magnet may be illustrated
in numbers of other ways. Thus, pivot a disk at its

centre, and arrange some light contact to touch its
edge, either at one point or all round, it matters not ;
then supply a current to disk from centre to circumference, and bring a bar-magnet near it along its axis,
or, better, two bar-magnets, with opposite poles one

FIG. 18.—Pivoted disk with radial current, revolving in a magnetic field and winding
up a weight.   The current is supplied to the axle by screw A, and leaves the rim
by mercury trough M.   The same apparatus obviously serves to demonstrate
currents induced by motion ; both directly and by the damping effect.

on each side, near the contact place of the rim ; the
disk at once begins to rotate (Figs. 18 and 19).

Instead of a disk one may use a single radius of it,
viz. a pivoted arm (Fig. 20) dipping into a circular
trough of mercury ; or we may use a light sphere
rolling on two concentric circular lines of railway

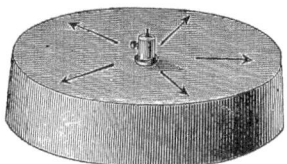

FIG. 19.—Another pivoted disk with flange to dip into liquid so as to make contact all round its rim. It rotates when a magnet is brought above or below ; or even in the field of the earth.

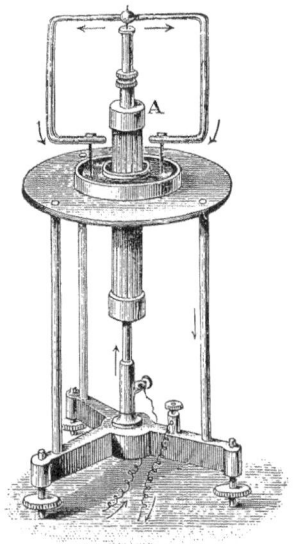

FIG. 20.—A couple of radii of the above disk provided with points to dip into mercury, and rotating constantly under the influence of the steel magnet A.

(Gore's arrangement, Fig. 21).[1]   In every case rotation
begins as soon as a magnet is brought near.

72. Nor is the revolving action confined to metallic
conductors and to true conduction.   Liquids and

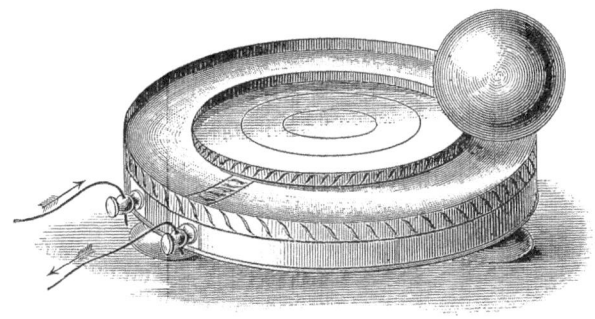

FIG. 21.—Gore's circular railway. The light spherical metal ball revolves round the
two concentric metal hoops or rails whenever it is made to convey a current
between them in a vertical magnetic field.

gases, although they convey electricity by something
of the nature of convection, are susceptible to rotation
in a precisely similar manner.

[1] This is not what Gore's railway is commonly used to illustrate, nor
is it the cause of the motion as observed by the inventor, or as described
in Tyndall's *Heat*.   Ordinarily the ball moves by reason of an irregular
disturbance due to heat at its point of contact with the rails, and it is
mere accident which way it goes.   But, in so far as the earth's vertical
magnetic field is strong enough, it should exhibit a preference for one
direction over the other ; and if the field is strengthened by bringing the
south pole of a bar-magnet below the apparatus, true magnetic rotation
is bound to occur.   It may, however, be convenient to state that the
current's own lines of force are entirely powerless to cause motion in
this case.   An external field is essential.

To show the rotation of liquid conductors under the influence of a magnet, take a circular shallow trough of liquid, supply it with stout sheet copper electrodes at centre and circumference, and put the pole of a magnet below it. The liquid at once begins to rotate, and by using a magnet and current of fair strength can easily be made to whirl so fast as to fly

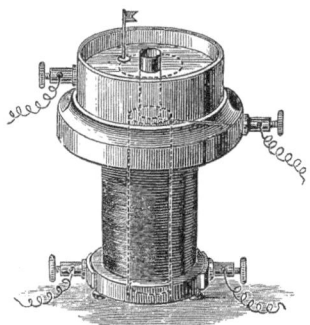

FIG. 22.—Rotation of a liquid disk conveying a radial current in a vertical magnetic field.

over the edge of the trough (Fig. 22).[1] The experiment is plainly the same as Fig. 19, except that a liquid disk is used in place of a solid one. Or, again,

[1] In practice it is most convenient to split a battery current between magnet and liquid : *i.e.* to connect them in parallel instead of in series. It is also well to make the smaller surface of copper the cathode ; because with intense currents (say 3 amperes per square centimetre) a crust of oxide forms on the anode which almost entirely stops the current by its resistance.

it may be considered the same as Fig. 21.   Reverse
the magnet, and the rotation is rapidly reversed.

Another method is to send a current along a jet of

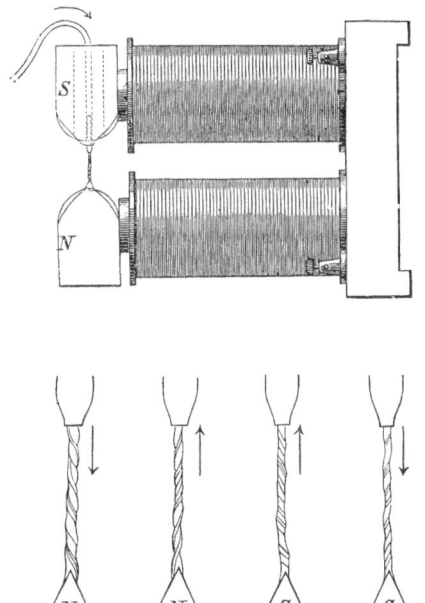

FIG. 23.—A falling stream of liquid conveying a current between two magnetic poles,
and being thereby twisted into a spiral. (Copied from a paper in *Phil. Mag.* by
Dr. Silvanus Thompson.)

mercury near a magnet and note the behaviour of the
jet.   It twists itself into a flat spiral as shown in
Fig. 23.

The rotation of a gas discharge is most commonly

illustrated by an arrangement like Fig. 24, where the terminals of the induction coil are connected to the rarefied gas respectively above one pole and round the middle of a magnetized bar. If the discharge can be got to concentrate itself principally down one side,

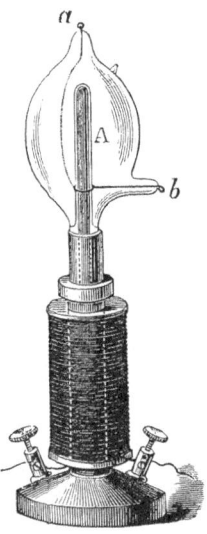

FIG. 24.—Induction coil discharge from *a* to *b* through rarefied gas, rotating round a glass-protected magnetized iron rod.

which it is not easy always to do (it seems to depend on the presence in the vacuum of some traces of foreign vapour, *e.g.* CS_2 vapour), the line of light so formed is seen to revolve.

*Action between a Magnet and an Electric Charge in
Relative Motion.*

73. From all this it is not to be doubted that a
charged pith ball moving in the neighbourhood of a
magnet is subject to the same action. There is no
known action between a magnet and a *stationary*
charged body, but directly either begins to move there
is an action between them tending to cause one to
rotate round the other. It is true that for ordinary
speeds of motion this force is extremely small ; but
still it is not to be doubted that if a shower of charged
pith balls or Lycopodium granules are dropped on to
a magnet pole, they will fall, not perfectly straight,
but slightly corkscrew fashion. And again, if a set
of charged particles were projected horizontally and
radially from the top of a magnet, their paths would
revolve like the beams of a lighthouse. And if by
any means their paths were kept straight, or deflected
the other way, they would exert on the magnet an
infinitesimal " couple " tending to make it spin on its
own axis.

Conversely, if a magnet were spun on its axis
rapidly by mechanical means, there is very little doubt
but that it would act on charged bodies in its neighbourhood, tending to make them move radially either
to or from it. This, however, is an experiment that

ought to be tried; and the easiest way of trying it
would be to suspend a sort of electrometer needle
electrified positive at one end and negative at the other,
near the spinning magnet, and to look for a trace of
deflection—to be reversed when the spin is reversed.
A magnet of varying strength might be easier to try
than a spinning one. (See §§ 114—116.)

Rotation of a Magnet by a Current.

74. The easiest way to show the actual rotation of
a magnet is to send a current half-way along it and
back outside. Thus, take a small, round, polished
steel bar-magnet with pointed ends, pivot it vertically,
and touch it steadily with two flakes or light pads of
tin-foil, one near either end and one near the middle;
supply a current by these contact pieces, and the
magnet spins with great rapidity. Reverse the current,
and it rotates the other way. Conversely, by producing the rotation mechanically a current will be
excited in a wire joining the two pieces of tin-foil
(Figs. 25, 26, and 27).

The two contacts may be made anywhere on the
magnet except symmetrically: if the two are equi-
distant from the middle, no effect will be produced.
The nearer one is to the middle and the other to
the end the stronger the effect; stronger still if

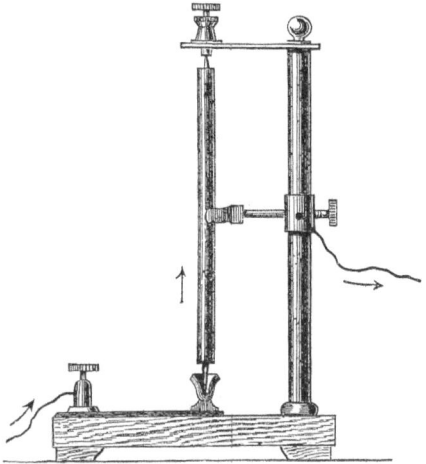

FIG. 25.—Round bright steel bar-magnet pivoted at its ends, spinning rapidly on its axis under the influence of a current supplied to either the bottom or top pivot, or both, and removed near the middle by a scrap of tin-foil lightly touching it.

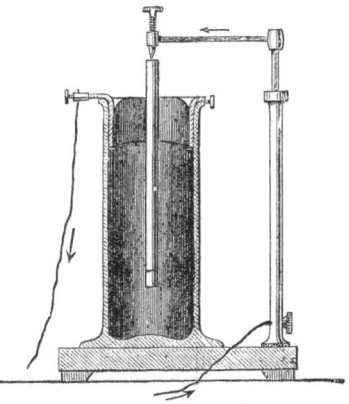

FIG. 26.—Another mode of exhibiting the same thing as Fig. 25. The magnet is loaded so as to float upright in mercury.

L

one of the contacts is either at or beyond the end.

75. The customary or Faraday plan of exhibiting the effect depicted in Fig. 25, with a mercury ring trough round the magnet into which a projecting wire carried by the magnet dips, is not quite so simple and obvious a method as Fig. 25 ; neither is it so effective unless

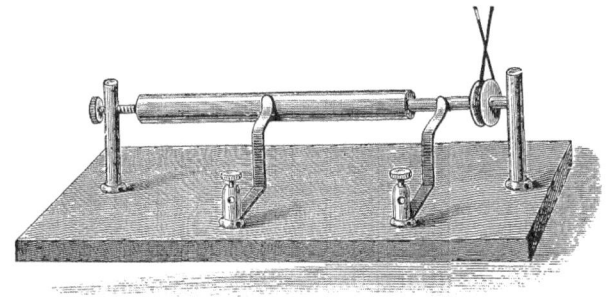

FIG. 27.—The converse of Fig. 25.   Spinning the magnet mechanically gives a current between two springs, one touching it near or beyond either end, the other touching it near the middle.

the ring trough fits the magnet pretty closely.   The arrangement in Fig. 25, where the contact is made actually on the surface of the magnet, gives the theoretically greatest force.

Many more variations of the experiment could be shown, but these are typical ones, and will suffice. They all call attention to the fact that a magnet, considered electrically, is a rotatory phenomenon.

CHAPTER VIII.

NATURE OF MAGNETISM.

Ampère's Theory.

76. THE idea that magnetism is nothing more nor less than a whirl of electricity is no new one—it is as old as Ampère. Perceiving that a magnet could be imitated by an electric whirl, he made the hypothesis that an electric whirl existed in every magnet and was the cause of its properties. Not of course that a steel magnet contains an electric current circulating round and round it, as an electro-magnet has ; nothing is more certain than the fact that a magnet is not magnetized as a whole, but that each particle of it is magnetized, and that the actual magnet is merely an assemblage of polarized particles. The old and familar experiment of breaking a magnet into pieces proves this. Each particle or molecule of the bar must have its circulating electric current, and then the properties of the whole are explained.

L 2

There is only one little difficulty which suggests itself in Ampère's theory—How are these molecular currents maintained? Long ago a similar difficulty was felt in astronomy—What maintains the motions of the planets? Spirits, vortices, and other contrivances were invented to keep them going.

But in the light of Galileo's mechanics the difficulty vanishes. Things continue in motion of themselves until they are stopped. Postulate no resistance, and motion is essentially perpetual.

What stops an ordinary current? Resistance. Start a current in a curtain-ring, by any means, and leave it alone. It will run its energy down into heat in the space of half a second or so. But if the metal conducted infinitely well there would be no such dissipation of energy, and the current would be permanent.

In a metal rod, electricity has to pass from atom to atom, and it meets with resistance in so doing; but who is to say that the atoms themselves do not conduct perfectly? They are known to have various infinite properties already; they are infinitely elastic, for instance. Pack up a box of gas in cotton-wool for a century, and see whether it has got any cooler. The experiment, if practicable, should be tried; but our present experience warrants us in assuming no loss of motion among colliding atoms until the contrary has been definitely proved by experiment. To all intents and purposes *certainly* atoms are infinitely elastic; why should they not also be infinitely conducting? Why should dissipation of energy occur in respect of an electric current circulating wholly inside an atom? There is no known reason why it should. There are many analogies against it.

How did these currents originate? We may as well ask, How did any of their properties originate? How did their motion originate? These questions are unanswerable. Suffice it for us, there they are. The atoms of a particular substance—iron for instance, or zinc—have an electric whirl of certain strength circulating in them as one of their specific physical properties.

This much is certain, that the Ampèrian currents are not producible by magnetic experiments. When a piece of steel or iron is magnetized, the act of magnetization is not an excitation of Ampèrian current in each molecule—is not in any sense a magnetization of each molecule. The molecules were all fully magnetized to begin with : the act of magnetization consists merely in facing them round so as to look mainly one way—in polarizing them, in fact. This was proved by Beetz long ago : I will not stop to explain it further, but will refer students to Maxwell (vol. ii. chap. vi.)

Ampère's Theory extended by Weber to explain Diamagnetism also.

77. Let us see how far we have got. We have made the following assertions :—

(1) That a magnet consists of an assemblage of polarized molecules.

(2) That these molecules are each of them permanent magnets, whether the substance be in its ordinary or in its magnetized condition, and that the act of magnetization consists in turning them round so as to face more or less one way.

(3) That when all the molecules are faced in the same direction the substance is magnetically completely saturated.

(4) That if each molecule of a definite substance contains an electric current of definite strength circulating in a channel of infinite conductivity the magnetic behaviour of the substance is completely explained.

But now, supposing all this granted, how comes it that the molecular currents are not capable of being generated by magnetic induction ? And if we cannot excite them, are we able to vary their strength ?

78. The answer to these questions is included in the following propositions, which I will now for convenience state, and then proceed to explain and justify.

(5) If a substance possessing these molecular currents be immersed in a magnetic field, all those molecules which are able to turn and look along the lines of force in the right direction will have their currents weakened ; but on withdrawal from the field they will regain their normal strength.

(6) If the currents naturally flowing in conducting channels be feeble or *nil*, the act of immersion of the substance in a magnetic field will reverse them or excite *opposite* currents, which will last so long as the body remains in the field, but will be destroyed by its removal.

(7) The same thing will happen whatever the strength of the natural molecular currents, provided the molecules are completely fixed and unable to face round under the influence of the field.

(8) The molecular currents so magnetically induced are sufficient to explain the phenomena of *diamagnetism.*

79. Let us first just recall to mind the well-known elementary facts of current induction. A conducting circuit, such as a ring or a coil of wire, suddenly brought near a current-conveying coil or a magnet, has a momentary current induced in it in the opposite direction to the inducing current—in other words, such as to cause momentary repulsion between the two. So long as it remains steady, nothing further happens ; but on withdrawing it another momentary

current is induced in it in the contrary direction to that first excited. The shortest way of expressing the facts quite generally is to say, that, while from any cause the magnetic field through a conductor is increasing in strength, a momentary current is excited in it tending to drive it out of the field ; and that, whenever the magnetic field decreases again to its old value, a reverse flow of precisely the same quantity of electricity occurs. Fig. 28 shows a mode of illustrating these facts. A copper disk is supported at the end of a torsion arm and brought close to the face of an unexcited bar electro-magnet. On exciting the magnet the disk is driven violently away : to be sucked back again, however, whenever the magnetism ceases.

80. Now, why are all these effects so momentary? What makes the induced current cease so soon after excitation ? Nothing but dissipation of energy : only the friction of imperfect conductivity. There is nothing to maintain the current, it meets with resistance in its flow through the metal, and so it soon stops.

But in a perfect conductor like a molecule no such dissipation would occur. Electricity in such a body will obey the first law of motion, and continue to flow till stopped. Destroying the magnetic field will stop an induced molecular current, but nothing else will stop it. Hence it follows that the repulsion

experienced by a molecule is no transitory effect like
that in Fig. 28, but is as permanent as the magnetic
field which excites and exhibits it.

Thus, then, a body whose molecules are perfectly
conducting, but without specific current circulating in
them, will behave diamagnetically, *i.e.* will move away

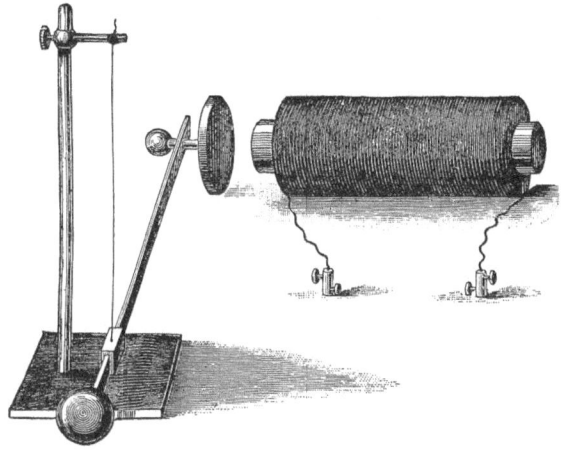

FIG. 28.—Stout disk of copper supported on a horizontal arm near one pole of a bar
electro-magnet.   The disk is repelled while the magnet is being excited, and is
attracted while the magnetism is being destroyed.

from strong parts of the field towards weak ones, and
thereby set its length equatorially, just as bismuth is
known to do.

Whether this be the true explanation of diamagnetism or not, it is at least a possible one. It seems to me
extremely probable. It is known as Weber's theory.

It does not necessarily follow that the specific molecular currents of a diamagnetic substance are really *nil;* all that is needful is that they shall be weaker than those induced by an ordinary magnetic field. By using an extremely weak field, however, the specific currents need not be quite neutralized, and in such a field the body ought to behave as a very feebly magnetic substance. Such an effect has been looked for.[1]

81. One loop-hole there is, however, viz. that every molecule may be so jammed as to be unable to turn round, and such a substance could hardly exhibit any noticeable magnetic properties. The molecules would have got themselves into a state of minimum potential energy, and if jammed therein nothing could be got out of them. The induced currents of diamagnetism would be superposed upon them just as if no initial molecular currents existed. By varying the temperature of such a substance, however, one might expect to alter its molecular arrangement, and so develop magnetic properties in it, just as electrical properties are developed in crystals like tourmaline by heat or by cold.

We are now able clearly to appreciate this much— that the molecular currents needful to explain magnetism are not conceivably excited by the act of magnetization, for they are in the wrong direction. *Induced* molecular currents will be such as to cause

[1] See *Nature,* vol. xxxiii. p. 484.

repulsion : those which cause attraction must have existed there before, and be merely rotated into fresh positions by the magnetizing force.   An intense magnetic field will weaken them, and thus tend to render a magnetic substance less magnetic.

Function of the Iron in a Magnet.   Two Modes of expressing it.

82. We can now explain the function of iron, or other magnetic substance, in strengthening a magnetic field.   Take a circular coil of wire, Fig. 29, and send a current round it : there is a certain field—a certain number of lines of force—between its faces.   Fill the coil with iron, so as to make it a common electromagnet, and the strength of the field is greatly increased.   Why ?   The common mode of statement likens the magnetic circuit to a voltaic circuit ; there is a certain magneto-motive force, and a certain resistance, or, as Mr. Heaviside preferably calls it, " reluctance " : the quotient gives the resulting magnetic induction, or total number of lines of force.   Iron is more permeable than air—say, 3000 times more permeable—and accordingly the resistance of the iron part of the circuit is almost negligible in comparison with that of the air-gap between the poles.   Thus a good approximation to the total intensity of field is obtained by dividing the magneto-motive force by the width of the air-gap ; or more completely and generally

by treating the varying material and section of a magnetic circuit just as the varying material and section of a voltaic circuit is treated, and so obtaining its total resistance.   Iron is thus to be regarded as a magnetic conductor between 100 and 10,000 times better than air.   Its specific magnetic conductivity or inductivity, or, as it is more usually called (after Thomson), permeability, is measured by the ratio of

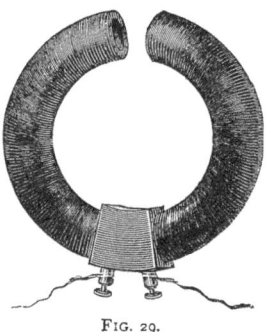

FIG. 29.

the magnetization produced to the magnetizing force applied, and is generally denoted by the symbol μ.[1]

83.   This mode of regarding the case is undoubtedly simple and convenient, but it is not the fundamental mode.   If we look at it less with a view to practical simplicity than with the aim of seeing what is really going on, we shall express it thus:—

Before the iron was inserted in the coil there were a certain number of circular lines of force inside it

[1] See Appendix (b).

due to the current alone. A piece of common iron, although full of polarized molecules, has no external or serviceable lines of force : they are all shut up, as it were, into little closed circuits inside the iron. But directly the iron finds itself in a magnetic field some of these open out, a chain of polarized molecules is formed, and the lines due to its molecular currents add themselves to those belonging to the current of the magnetizing helix.

Thus our ring electro-magnet has now not only its own old lines of force, but a great many of those belonging to the iron which have sympathetically laid themselves alongside the first.

Parenthetically we may make the following remark. So long as the iron adds some 3000 lines of its own (more or less according to the quality of the iron) for every one otherwise excited in the field, so long it has its maximum permeability : it is infinitely far from saturation.   But after a certain call upon it, it begins to show signs of poverty, and ultimately may refuse to add any lines of its own at all ; it is then said to be completely saturated : its permeability is then just as if it were air.   The permeability of iron is an extremely indefinite quantity.   Not only does it vary with the same piece as it nears saturation, but it is exceedingly different for different specimens.   Thus some manganese-steel exists with a permeability one and a half times that of air, or only about as much as zinc, while Ewing finds some iron with a permeability as high as 20,000 under shaking.

The end result of either mode of regarding the matter is of course the same—the lines of force between the poles are increased in number by the presence of iron ; but whereas, in the first-mentioned mode of treatment, the fact of permeability had to be accepted unexplained, in the second nothing is unexplained except the fundamental facts of the subject, such as the reason why currents tend to set themselves with their axes parallel, and other matters of that sort.

Permanent Magnetism.

84. There is one curious effect of introducing iron or other solid magnetic medium into a magnetic field which must not be overlooked, This effect depends on the solidity of the substance, i.e. the fixedness or stiffness of its molecules. In a fluid the molecules are free to take up any fresh arrangement with ease : there is no set arrangement in the internal structure of a fluid, any more than there is a definiteness in its external shape. But with a solid it is different: its molecules once set into any position tend to remain more or less in that position ; the substance may be elastic to a certain extent, but after large disturbances there will always be a certain amount of permanent

set. Hence it is that solid bodies have a definite shape, which it requires force to change ; hence it is also that their molecules are able to crystallize into geometrical patterns.

Now, since the act of magnetization consists in making a number of already polarized molecules face round more or less in one direction, it follows that solid magnetic substances will behave differently from fluid ones. In fluid media the magnetized arrangement can only be maintained by a continuous exertion of magnetizing force ; and directly this is withdrawn the molecules will quickly take up their old higgledy-piggledy arrangement of minimum energy, and all trace of magnetization will cease. They will be perfectly easy to magnetize, and they will automatically demagnetize themselves. But with solids it is otherwise. The molecules if set in their magnetized position by only a feeble force will spring back almost completely when the magnetizing force is removed ; but if they have been arranged by a force of some violence, the spring back will be only partial, and a permanent set will remain. The spring-back portion of the whole arrangement is called temporary magnetism, the set portion is called permanent magnetism. The difference can be illus-trated by bending a bit of tin plate or paper nearly double, and then letting it go.

Substances differ greatly in their power of thus

retaining magnetization, and, as is well known, steel has the property well developed ; but all substances exhibit it more or less.[1] Moreover, many substances can retain a little of the set if they are left carefully undisturbed, but they lose it if shaken or heated : sometimes even if gently touched. A long thin bar of soft iron is most instructive in this respect. It can be easily magnetized by the earth's magnetism if held vertically and struck with a finger. If then inverted slowly and cautiously, it will retain nearly the whole of the induced magnetism ; but if struck again, or even if the fingers are shuffled on it (so sensitive some bars are), the whole is immediately reversed. Soft iron can, in fact, retain enormously more magnetism than steel can, but it retains it in a very feeble and loose manner. Its magnetism can only be styled sub-permanent.

85. A short thick bar can retain much less magnetism than a long thin one ; in fact, if a stout bar is made of the softest iron, it can retain hardly any. A piece of iron shaped like Fig. 29 would have a much better chance of retaining its magnetism, and if the gap were closed by another piece of iron or keeper, it would retain it very well ; while if the last trace of air-gap, the air-films between keeper and magnet, be abolished by making the whole one welded ring, then its magnetism is retained almost perfectly.

[1] See a letter in *Nature*, vol. xxxiii. p.'484.

There is some demagnetizing force even in this case, for a fluid magnetized as a ring would not remain magnetized, and I find that tapping or beating such a ring does appreciably weaken its magnetization, but the demagnetizing force is very small compared with what it is when there is an air-gap.

Hence we learn that the specially demagnetizing portion of a magnetic current is the fluid portion—the air portion ; and the greater the proportion of the fluid portion to the whole, the more easily is demagnetization accomplished. Fluids having no power of their own for retaining magnetism, if lines of force are forcibly maintained in air or other fluid in the neighbourhood of a solid magnet, all the strain of upholding not only its own magnetism, but all the rest of the magnetism in the field—the strain of keeping the molecules faced round in opposition to their mutual restoring forces—has to be thrown upon the viscosity and retentivity of the solid.

86. All the known facts of magnetism have had new life and interest put into them lately by the researches of Ewing. The long-known fact that solid substances store up in their structure any previous arrangement of their molecules, so that traces of the effect are recognizable long after the cause of the effect has been withdrawn, is not indeed by any means confined to magnetism ; it is a general property of solids, and constitutes a considerable difficulty in dealing with

M

them theoretically. The properties of all fluids, whether liquids or gases, depend upon their state at the moment, and upon nothing else : not at all upon how they reached that state, or upon what has happened to them in past times. Hydrogen at 0° C. and 76 centimetres pressure is a perfectly definite substance. Water at 50° C. and one atmosphere pressure, is again a quite complete specification. And the same is very nearly true of some crystalline solids. Quartz or ice at given temperature and pressure is generally considered quite a definite statement, though perhaps it is not so exactly definite as we imagine. But glass, or steel, or copper, at a specified temperature, is by no means a definite substance. If it has been cooled down to that temperature it will not be the same as if it had been warmed up to it. We must be told whether it has been hardened, or tempered, or annealed, and so on. The properties of a solid body depend on its past history as well as on its present state.

All this is pre-eminently true with magnetization. To understand completely the behaviour of a magnet we must not only know its present state, but we must know how it got to that state. A piece of steel once magnetized and then demagnetized is not in the same condition as if it had never been in a magnetic field : unless, indeed, it has been melted and made afresh.

This much, however, must be granted, that if *every-thing* were known about the instantaneous state of a

body there would be no need to go back upon its past
history : it might even be possible to deduce some of
its past history from its present state.   But it is precisely because a knowledge of the position and relation
of every individual molecule is impossible, and because
we have to put up with a few salient features of
information, that an inquiry into past history is
necessary.   A few salient features are sufficient in the
case of fluids : they are, in general, not sufficient in
the case of solids.

I have laid stress upon this matter because it is an
important general distinction between states which are
self-contained, so to speak, and states which are led up to.

87. A further detail of the distinction is that, in
solids, a direct and return series of changes are not
usually the same ; a precisely inverse cause does not
precisely invert the effect.   Take a body from one
self-contained state to another.   Then, whenever you
reverse the series of operations which brought it there,
it will return by the same path to its previous condition, and everything will be as it had been, and no
work need have been done on the whole.   Not so with
the led-up-to states.   Magnetize a piece of steel by
one series of operations, and then perform the same
operations in reverse order, it will not return by the
same path at all, nor will it return to its original condition.   Continue the process of magnetization and
reverse magnetization several times, and you may at

length succeed in getting the body to go through a cycle of changes, at least approximately. But it will go by one path and return by another.

Now, when a body of any kind is taken from a state A to a state B by one path, and back by some other path, as steam is for instance in a steam-engine, the result is always that some work is either done by the substance, or has to be done upon it, in performing the cycle. The return by a different path is optional in the case of steam, and accordingly work may or may not be done on it; but in the case of a magnetized solid it is not optional. The ascending curve of increasing magnetism, and the descending curve of decreasing magnetism, do not coincide, and cannot be made to coincide. Consequently, whenever a piece of iron is taken round a cycle of magnetic changes, some work is necessarily done.

This work, in general, results in a production of heat, and accordingly a piece of iron magnetized and demagnetized successively in rapid succession gets slightly warm. This direct heating effect is, however, very small, and is usually inappreciable in practice.[1]

All this behaviour of iron and other substances with regard to magnetism is called by Ewing hysterēsis.

---

[1] The large indirect heating by induced currents—the so-called Foucault effect—is too familiarly known to need any other statement than that it is quite distinct from what we are here discussing.

Electrical Momentum once more.

88. There is just one point which I must stop here to call attention to. The theories of magnetism and diamagnetism, which I have given according to Ampère, Weber, and Maxwell, require as their foundation that in a perfect conductor electricity shall obey the first law of motion—shall continue to flow until stopped by force. But the property of matter which enables it to do this is called *inertia;* the law is called the law of inertia ; and anything which behaves in this way must be granted to possess inertia.

It would not do to deduce so important a fact from a yet unverified theory ; but at least one must notice that momentum is essentially involved in Ampère's theory of magnetism. It is the only theory of magnetism yet formulated, and it breaks down unless electricity possesses inertia.

Nevertheless it is a fact that an electro-magnet does not behave in the least like a fly-wheel or spinning-top : there is no momentum mechanically discoverable (§ 39). Supposing this should turn out to be strictly and finally true, we must admit that a molecular electric current consists of two equal opposite streams of the two kinds of electricity : one must begin to regard negative electricity not as merely the negation or defect of positive, but as a separate entity. Its relation to positive may turn out to be something more like that of sodium to chlorine than that of cold to heat.

89. I said that no effect due to electric inertia was *mechanically* discoverable ; and on the hypothesis that an electric current consists of a pair of equal opposite currents of positive and negative electricity respectively this is very natural. Think of a couple of india-rubber pipes tied together so as to form a double tube, and through each propel a current of water, one in an opposite direction to the other. The double current has no gyrostatic properties, and the only way the water can exhibit momentum is by its resistance to change of velocity, like the "extra-current" effects in electricity.[1]

So long as one considered the flow of electricity in ordinary conductors, we could partially avoid the question of inertia by considering it urged forward at every point with a force sufficient to overcome the resistance there and no more ; but though this ex-

[1] When these articles originally appeared in *Nature* I made a serious hydraulic mistake about this place, saying that, just as a loop of very light flexible thinly-covered stranded wire or gold-thread, thrown at random on a glass slab and a strong current passed through it, tended to round off its sharp corners, open out its tangled loops, and do its best to become a perfect circle, so would a stream of water behave, as regards kinks and bends and curves in its pipe. Prof. Rücker was kind enough to call my attention to this mistake. It is, indeed, well known that not the slightest effect on the shape of a flexible pipe is produced by a stream of water in it, *so long as its ends are fixed :* the stream rather confers a kind of rigidity on the pipe.

plained the shape of the stream-lines (p. 103) yet it
did not suffice to render clear the phenomena of self-
induction—the lag of the interior electricity in a wire
behind the outside until definitely pushed (§§ 43-48) ;
still less does it explain its temporary persistence in
motion after the pushing force has ceased.

But, now that we are dealing with perfect conductors with no pushing force at all, the persistence of
molecular currents without inertia, or an equivalent
property so like it as to be rightly called by the same
name at present, becomes inexplicable. True, the
molecular currents are as yet an hypothesis ; and that
is the only loop-hole out of a definite conclusion
(§§ 98 and 185).

CHAPTER IX.

STRUCTURE OF A MAGNETIC FIELD.

90. LET us now pass in review the various facts and experiences which have led us to a dual view of electricity ; a kind of two-fluid theory, but in a very modified form.

First, there are the old experiments which vaguely suggest the separate existence of negative electricity, such as :—

(1) The wind from a point whether positive or negative ; so that a candle gets blown always away from it, whether the point be on the prime conductor and the candle held in the hand, or whether the point be held in the hand and presented to the candle or prime conductor ; so, also, that a point whirligig turns the same way, whether supported on the prime conductor, or whether attached to the earth and placed near it.

(2) Phenomena connected with the spark discharge,

such as Wheatstone's old experiment on what he called the velocity of electricity, with the three pair of knobs ; and the double burr produced in cardboard when pierced with a spark, suggesting that something has pierced it both ways at once.

Then there are the more recently observed facts ; as, for instance :—

(3) The fact that an electrostatic strain scarcely affects the volume of a dielectric ; thereby at once suggesting something of the nature of a shearing or distorting stress, which alters shape but not size ; a displacement of positive outwards and simultaneous negative inwards (§ 13).

(4) The facts of electrolysis, and the double procession of atoms past each other in opposite directions.

(5) The phenomena of self-induction, and the behaviour of a thick wire to an alternating current. The delay also in magnetizing iron, and especially the possibility of permanent magnetism ; combined with

(6) The absence of momentum in an electric current, or moment of momentum in an electromagnet, as tested by all mechanical means yet tried.

I admit at once that many of these are mere superficial suggestions which may hardly bear examination and criticism. Only (3), (4), (5), and (6) can be at all seriously appealed to ; but (5) and (6), in conjunction, seem to me to afford a sort of provisional and hypothetical proof, which (3) greatly strengthens.

At this point we must for the present again leave the question. We return to it in § 118 and § 155.

Representation of a Magnetic Field.

91. The disturbance called magnetism, which we have shown in Chap. VII. to be something of the nature of a spin—a rotation about an axis—is conspicuously not limited to the steel or iron of the magnet : it spreads out through all adjacent space, and constitutes what is called the magnetic field. A map of the field is afforded by the use of iron filings, which cling end to end and point out the direction of the force at every point (Fig. 33).

These lines of force so mapped are to be regarded as the axes of molecular whirls (Fig. 30). They are continuous with similar lines in the substance of the steel, and every line really forms a closed curve, of which a portion is in the steel and a portion in the air. In a wire helix, such as Figs. 16 or 29, the lines are wholly in the air, but in one part of their course they thread the helix, and in another part they spread out more or less between its faces.

But according to Ampère's theory of molecular currents there is no *essential* difference between such a helix and a steel magnet; directly the currents in the molecules of the magnet are considered, everything resolves itself into chains of molecular currents, threading themselves along a common closed curve or axis. Each atom, whether in the steel or in the air, is the seat of a whirl of electricity, more or less faced round

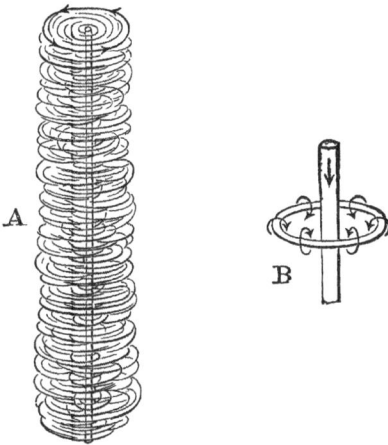

FIG. 30.—A, an element of a magnetic line of force with the electric whirl round it; B, a bit of an electric circuit with one of its magnetic lines of force shown round it, and the electric whirl round this : each magnetic line of force round a current being an electric vortex ring. Compare Fig. 39.

so as on the average to have its plane at right angles to the lines of force. The simplest plan of avoiding having to consider those only partially faced round, is to imagine the whole number divided into a set which face accurately in the right direction, and a set which look any way at perfect random ; and to neglect this latter set.

92. Well now try and picture a chain of whirls like beads spinning on a wire threading them all, and think of the effect of a material fluid thus rotating. Obviously it would tend to whirl itself fatter, and to shorten its length. An assemblage of such parallel straight whirls would thus squeeze each other later-ally, or cause a lateral pressure, and would tend to drag their free ends together, causing a longitudinal tension.

Such whirls cannot in truth have free ends except at the boundary of a medium—as at the free surface of a liquid. Magnetic whirls are in reality all closed curves ; but inasmuch as part of them may be in a mobile fluid like air, and part of them in a solid like iron or steel, it is convenient to distinguish between their two portions ; and one may think of the air whirls alone, as reaching from one piece of iron to another and by their shortening tendency or centrifugal force pulling the two pieces together.

The arrangement shown in Fig. 31 illustrates the kind of force exerted by a spinning elastic framework, along and perpendicular to its axis of rotation.

One can easily see this effect of a whirl in a tea-cup or inverted bell-jar full of liquid. Stir it vigorously, and leave it. It presses against the walls harder than before, so that if they were elastic they would bulge out with the lateral pressure ; and it sucks down the top or free end of its axis of rotation, producing quite a depression or hollow against the force of gravity Or, as a more striking illustration, make the apparatus sketched in Fig. 32.

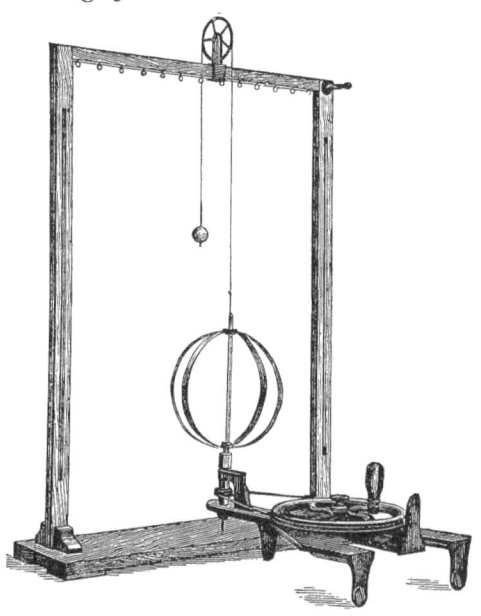

FIG. 31.—A "shape of the earth" model which, when whirled, exerts a tension along its axis, pulling up the weight attached to it, and a pressure at right angles, by reason of its bulging out.

Two circular boards joined by a short wide elastic tube or drum : a weight hung to the lower board, the top board hung from a horizontal whirling table, the drum filled with water, and the whole spun round.

The weight is raised by the longitudinal tension ; the sides bulge out with the lateral pressure.

There is no need for the whole vessel to rotate. If the liquid inside rotates, the same effect is produced.

93. Imagine now a medium composed of a multitude of such cells with rotating liquid inside : let the cells be either very long, or else be joined end to

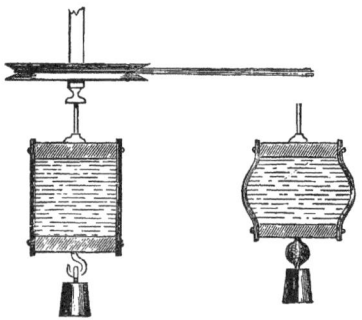

FIG. 32.—An elastic-walled cylindrical vessel full of liquid hanging from a whirling table, and, by reason of centrifugal force, raising a weight and bulging out laterally when spun, thereby illustrating a tension along the axis of rotation and a pressure in every perpendicular direction.

end so as to make a chain—a series of chains side by side—and you have a picture of a magnetic medium traversed by a field of force.   End-boundaries of the field will be dragged together, thus representing magnetic attraction ; while, sideways, the lines of force (axes of whirl) squeeze each other apart, thus illustrating repulsion.   This is Clerk-Maxwell's view of

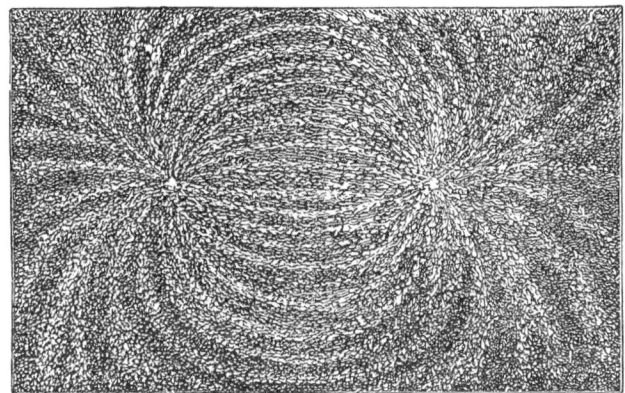

Attraction.

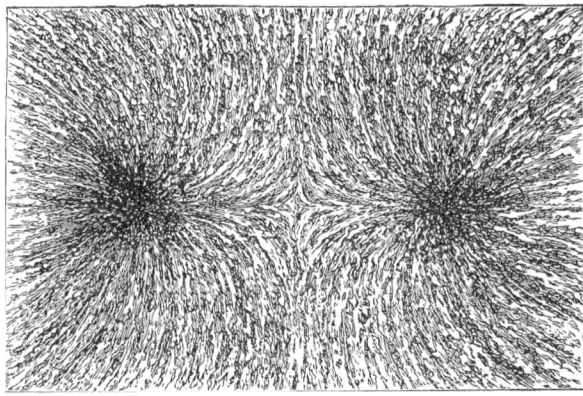

Repulsion.

FIG. 33.—Attraction and Repulsion. The tension along the lines of force or axes of rotation drags the one pair of poles together; and the pressure in directions perpendicular to the axis of rotation, due to the centrifugal force of the whirls, drives the other pair apart.

an electro-magnetic medium, and of the mode in which magnetic stress, and magnetic attractions and repulsions between bodies, arise.

Wherever lines of force reach across from one body to another, those bodies are dragged together as if pulled by so many elastics (Fig. 33) ; but wherever lines of force from one body present their *sides* to those proceeding from another body, then those bodies are driven apart.

CHAPTER X.

MECHANICAL MODELS OF A MAGNETIC FIELD.

First Representation of the Field due to a Current.

94. RETURN now to the consideration of a simp
circuit, or, say, a linear conductor, and start a current
through it ; how are we to picture the rise of the lines
of force in the medium ? how shall we represent the
spread of magnetic induction ? First think of the
current as exciting the field (instead of the field as
exciting the current, which may be the truer plan
ultimately).

If we can think of electricity in the several molecules of the insulating medium connected like so
many cog-wheels gearing into one another and also
into those of the metal, it is easy to picture a sideways spread of rotation brought about by the current,
just as a moving rack will rotate a set of pinions
gearing into it and into each other (Fig. 34). But

N

then half the wheels will be rotating one way and half the other way, which is not exactly right.

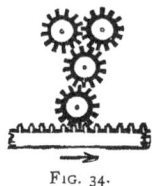

FIG. 34.

How is it possible for a set of parallel whirls to be all rotating in the same direction ?

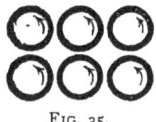

FIG. 35.

If there is any sort of connection between them they will stop each other, because they are moving in opposite directions at their nearest points ; and yet, if there is no connection, how can the whirl spread through the field ?

Well, return to the old models by which we endeavoured to explain electrostatics, and think whether they will help us if we proceed to superpose upon them a magnetic whirl in addition to the properties they already possess.   Looking at Figs. 5, 6, and 7A, we remember we were led to picture atoms and electricity like beads threaded on a cord. And these cords had to represent, alternately, positive and negative electricity, which always got displaced in different directions (see § 90).

We are forced to a similar sort of notion in respect of the wheels at present under discussion ; in order that they may co-operate properly, they must represent positive and negative electricity alternately.   If

FIG. 36.—Rows of cells alternately positive and negative, geared together free to turn about fixed axles.

they *then* rotate alternately in opposite directions, all is well, and the electrical circulation or rotation in the field is all in one direction.   Each wheel gears into and turns the next, and so the spin gets propagated right away through the medium, at a speed depending on the elasticity and density concerned in such disturbances.

It is not convenient at the present stage to ask the

N 2

question whether the wheels represent atoms of matter or merely electricity. It may be that each atom is electrostatically charged and itself rotates, in which case it would carry its charge round with it, and thereby constitute the desired molecular current. The apparent inertia of electricity would thus be explained simply enough, as really the inertia of the spinning atoms themselves ; and the absence of any

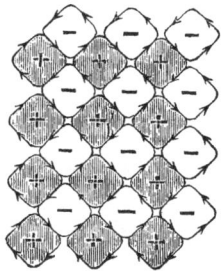

Fig. 37.—Section of a magnetic field perpendicular to the lines of force   Another mode of drawing Fig. 36.

moment of momentum in an electro-magnet as tested mechanically would be equally explained by the simultaneous opposite rotation of adjacent atoms. A question may arise as to why the opposite molecules should have exactly equal opposite in-ertiæ, as they have, else a fluid magnetized medium would bodily rotate ; and there may be other difficulties connected with a bodily rotation of electrostatically charged molecules : it is merely a possibility upon which stress must not be laid till it has been proved able to bear it. For our present purpose a spin of the electricity inside each atom, or even independently of any atoms, is quite sufficient. Besides, since magnetic induction can spread through a vacuum quite easily, the wheel-work has to be largely independent of material atoms.

If any difficulty is felt concerning the void spaces in Fig. 36, it is only necessary to draw it like Fig. 37, which does every bit as well, and reduces the difficulty to any desired minimum.

Representation of an Electric Current.

95. Now notice that in a medium so constituted and magnetized—that is, with all the wheel-work revolving properly—there is nothing of the nature of an electric current proceeding in any direction whatever. For, at every point of contact of two wheels the positive and negative electricities are going at the same rate in the same direction ; and this is no current at all. Only when positive is going one way and negative going the opposite way, or standing still, or at least going at a *different* rate, can there be any advance of electricity, or anything of the nature of a current.

A current is nevertheless easily able to be

represented : for it only needs the wheels to gear imperfectly and to work with slip. At any such slipping-place the positive is going faster than the negative, or *vice versâ*, and so there is a current there. A line of slip among the wheels corresponds therefore to a linear current ; and, if one thinks of it, it is quite plain that such a line of slip must always have a closed contour. For, if only one wheel slip, then the circuit is limited to its circumference; if a row slip, then the direct and return circuit are on opposite sides of the row. But if a large area of any shape with no slip inside it is inclosed by a line of slip, then this gives us a circuit of any shape, but always closed. Understand : one is not here thinking of a current as analogous to a *locomotion* of the wheels—their axes may be quite stationary,—the slip contemplated is that of one *rim* on another.

Imagine all the wheels inside the empty contour of Fig. 38 to be rotating, the positive clockwise, the negative counter clockwise, and let all those outside the contour be either stationary or rotating at a different rate or in an opposite direction ; then the boundary of the inside region is a line of slip, along which the positive rims are all travelling clockwise and the negative rims the other way, and hence it represents a clockwise positive current round the inside of the empty contour.

But it may be said that the spin inside the contour,

if maintained, must sooner or later rotate the wheels outside as fast as themselves, and then all slip will cease. Yes, that is so, unless there is a complete breach of connection at the contour, as in Fig. 38

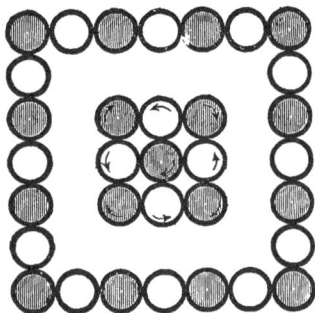

FIG. 38.—Diagram of a peripheral current partitioned off from surrounding medium by a perfect conductor, which transmits no motion, and therefore acts as a perfect magnetic screen. See also § 101, and Fig. 41. Shaded wheels stand for positive.

there is. If the outer region has any sort of connection with the inner one, the slip at its boundary can only be temporary, lasting during the era of acceleration.

Distinction between a Dielectric and a Metal, as affected by a spreading Magnetic Field.

96. In a dielectric the connection between the atoms is definite and perfect. If one rotates, the next must rotate too ; there is no slip between the geared

surfaces ; it is a case of cogged wheels like Fig. 36. A conduction-current is impossible.

But in a metallic conductor the gearing is imperfect ; it is a case of friction-gearing with more or less lubrication and slip, so that turning one wheel only starts the next gradually—it may be very quickly, but not instantaneously—and there is a motion of a positive rim incompletely compensated by an equal similar motion of a negative rim while getting up speed ; in other words, there is a momentary electric current, lasting till the wheels have fairly started.

In a perfect conductor the gearing is absent ; the lubrication is so perfect that all the atoms are quite free of one another, and accordingly a spin ceases to be transmitted into such a medium at all. The only possible current in a perfect conductor is a skin-deep phenomenon. (See also Chap. V. and § 104.)

A magnetized medium of whatever sort is thus to be regarded as full of spinning wheels, the positive rotating one way and the negative the other way. If the medium is not magnetized, but only magnetic —*i.e.* capable of being magnetized—it may be thought of either as having its wheels stationary, or as having them facing all ways at random ; the latter being probably the truer, the former the easier, representation, at least to begin with.

Whether the medium be conducting or insulating makes no difference to the general fact of spinning wheels inside it wherever lines of force penetrate it. But the wheels of a conductor are imperfectly cogged together ; and accordingly, in the variable stages of a magnetic field, while its spin is either increasing or decreasing, there is a very important distinction to be drawn between insulating and conducting matter. During the accelerating era conducting matter is full of slip, and a certain time elapses before a steady state is reached. A certain time may be necessary for the propagation of spin in a dielectric, but it is excessively short, and the process is unaccompanied by slip, only by slight distortion and recovery. (See §§ 103 and 159.)

97. As for strongly magnetic substances like iron, nickel, and cobalt, one must regard them as constituted in the same sort of way, but with wheels greatly more massive, or very much more numerous, or both. The quantity which we have called permeability in §§ 82, 83, and denoted by the symbol μ, may now be thought of as physically equivalent to a density of the magnetic medium ; so that substances with a large μ, like iron, have their magnetic mechanism or wheel-work exceedingly massive.

Phenomena connected with a varying Current. Nature of Self-induction.

98. Proceed now to think what happens in the region round a conductor in which a current is rising.

Without attempting a complete and satisfactory representation of what is going on, we can think of some mechanical arrangements which have some close analogy with electrical processes Chap. V.

Take first a system of wheel-work connected together and moved at some point by a rack. Attend to alternate wheels more especially, as representing

FIG. 39.—A provisional representation of a current surrounded by dielectric medium, either propelling or being propelled. Section through the wire.

positive electricity. The intermediate negative wheels are necessary for the transmission of the motion, and they also serve to neutralize all systematic advance of positive electricity in any one direction, except where slip occurs, but they need not otherwise be specially attended to.

Remember that every wheel is endowed with inertia, like a fly-wheel (§ 88).

Directly the rack begins to move, the wheels begin to rotate, and in a short time they will all be going full speed. Until they are so moving, the motion of the rack is opposed, not by friction or ordinary resistance, but by the inertia of the wheel-work. *This inertia represents what is called self-induction,* and the result of it is what has been called the " extra-current at make," or, more satisfactorily, the opposing E.M.F of electro-magnetic inertia or self-induction.

Having once started the rack, so long as it moves steadily forward, the wheel-work has no further effect upon it ; but, directly it tries to stop, it finds itself unable to stop dead without great violence : its motion is prolonged for a short time by the inertia of the wheel-work, and we have what is known as the " extra-current at break."

99. If the rack is for a moment taken to represent the advancing electricity in a copper wire, then the diagram may be regarded as a section of the complete field : the complete field being obtained from it by rotating the diagram round the axis of the rack. Imagining this done, we see that the axis of each wheel becomes prolonged into a circular core, and each wheel into a circular vortex ring surrounding the rack and rolling down it as it moves forward, as when a stick is pushed through a tight-fitting umbrella-ring held stationary (see Fig. 30, B).

As one goes further and further from the rack the

lengths of the vortex cores increase, but there is only a given amount of rotation to be shared among more and more stuff, hence it is not difficult to imagine the rate of spin diminishing as the distance increases, so that at a reasonable distance from the conductor the medium is scarcely disturbed.

100. To perceive how much rotation of the medium is associated with a given circuit, one must consider the shape of its contour—the position of the return current.

Take first a long narrow loop and send a current up one side and down the other. The rotations belonging to each are superposed, and though they agree in direction for the space inclosed by the loop, they oppose each other outside, and so there is barely any disturbance of the medium outside such a looped conductor ; very little dielectric is disturbed at all, and accordingly the inertia or self-induction is very small (see Fig. 40).

If the loop opens out so as to inclose an area, as the centrifugal force of the wheels will tend to make it do, then there is a greater amount of rotation, a greater moment of momentum inside it, and accordingly its self-induction is increased. The axis of every wheel is, however, continuous, and must return outside the loop : so the outside region is somewhat affected by rotation, but of a kind opposite to that inside.

101. Figs. 38 and 41 show the state of things for a

closed circuit conveying a current. The free space in Fig. 38 represents a perfect conductor, or perfect breach of connection. Along the inner boundary of this space positive electricity is seen streaming in the direction of the arrows, and it may be streaming quite independently on the outer boundary also, but

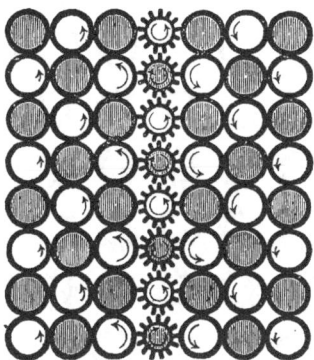

FIG. 40.—Diagram of a direct and return current close together, showing distribution of rotation and of slip in the thickness of the conductor, and in the dielectric between. The dielectric outside is very little disturbed. The smooth wheels represent a metallic conductor, the cog wheels represent a thin layer of dielectric between the direct and return part of the metallic circuit.
For two currents going in the same direction, see Fig. 44.

nothing happens in its interior—which is therefore not represented.

The corresponding portion in Fig. 41 is intended for an ordinary conductor, full of wheels capable of slip. And slip in this case is a continuous necessity, for the rotation on either side of the conductor is in opposite directions, so the atoms of the conductor have to

accommodate themselves as best they can to the
conditions ; some of them rotating one way, some the
other, and some along a certain neutral line of the

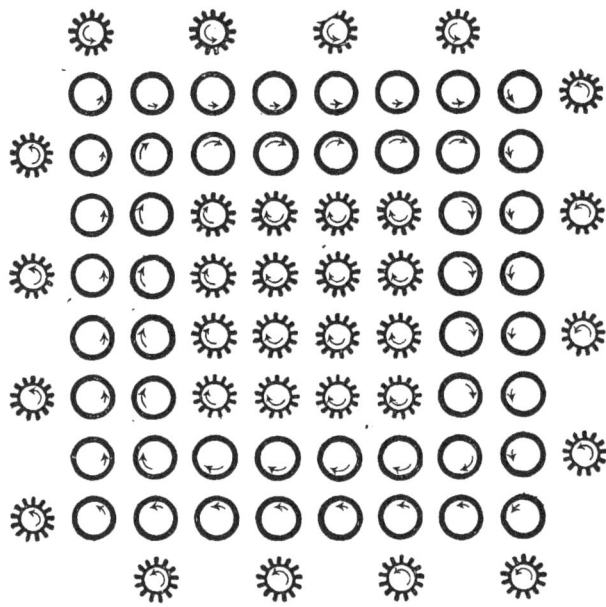

FIG. 41.—Diagram of simple conducting circuit like a galvanometer ring, with the
alternate connecting-wheels omitted. The same number of dielectric wheels are
drawn outside as inside, to indicate the fact that the total spin is equal inside
and out, though the outside is so spread out as to be much less intense. The
diagram shows a clockwise positive current flowing steadily round the ring ; with
the accompanying distribution of magnetism. Section in the plane of the ring.

conductor being stationary. If a conductor is straight
and infinitely long, the neutral line of no rotation is
in the middle. If it be a loop, the neutral line is

nearer the outside than the inside, because the rotation of the medium inside is the strongest. If the loop be shut up to nothing, the neutral line is its outer boundary or nearly so (Fig. 40). If, again, the circuit is wound round and round a ring, as string might be lapped upon a common curtain-ring to cover it, then the axes of whirl are wholly inclosed by the wire, and there is no rotation outside at all.

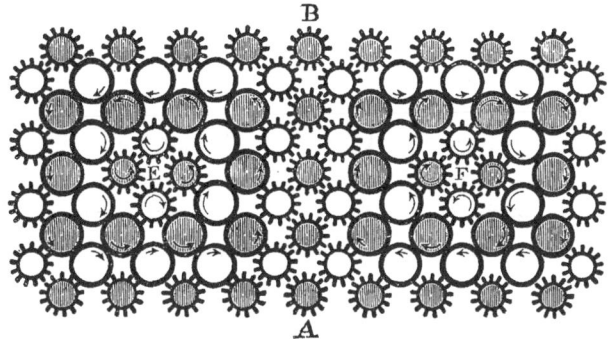

FIG. 42.—Section of a closed magnetic circuit, or electric vortex-ring, or hollow bent solenoid like Fig. 29, inclosing an anchor-ring air space ; the axis of the ring being A B, the sections of its core being E and F. The arrows indicate the intensity of the spin, *i.e.* of the magnetic field, which is a maximum at the middle of each section and nothing at all outside. If the core contains iron instead of air, its wheels have to be from 100 to 10,000 times as massive : slipping wheels if solid iron, cogged wheels if a bundle of fine varnished iron wires. Cf. Fig. 47.

Fig. 42 shows a section of this last-mentioned condition, and here the wheels of the dielectric outside are not rotating at all. The inside is revolving, it may be furiously, and so between the inner and outer layers of the conductor we have a great amount of slip and dissipation of energy.

102. The process of slip which we have depicted
goes on in all conductors conveying a current, whether
steady or variable, and in fact *is* the current. The
slip is necessarily accompanied by dissipation of
energy and production of heat : only in a perfect conductor can it occur without friction. In a steady current the slip is uniformly distributed throughout the
section of the conductor ; in the variable stages it is
unequally distributed, being then more concentrated
near the periphery of the wire (§ 43).

When a current is started in a wire, the outer
layers start first, and it gradually though very quickly
penetrates to the axis. Hence the lag or self-induction of a wire upon itself is greater as the wire is
thicker, and also as it is made of better conducting
substance. If it is of iron, the mass or number of the
wheels is so great that the lag is much increased, and
the spin of its outer layers is great enough to produce
the experimental effects discovered by Prof. Hughes.

One must never confuse the slip with the spin. Slip
is current, spin is magnetism. There is no spin at the
axis of a straight infinite wire conveying a current,
and it occurs in opposite directions as you recede
from the axis either way ; arranging itself in circular
vortex cores round the axis (Fig. 30, B). But the slip
is uniformly distributed all through the wire as soon
as the current has reached a steady state. The slip is
wholly in the direction of the wire. The axes of spin
are all at right angles to that direction.

CHAPTER XI.

MECHANICAL MODELS OF CURRENT INDUCTION.

Rise of Induced Current in a Secondary Circuit.

103. To study the way in which a magnetic field excited in any manner spreads itself into and through a conducting medium, look at Fig. 43, and suppose the region inside the contour A B C D to be an ordinary conducting region—that is, full of wheels imperfectly geared together, and capable of slip.

Directly the rack begins to move, all the wheels outside A B C D begin to rotate, and quickly get up full speed. The outer layer of wheels inside the contour likewise begins to rotate, but not at once; there is a slight delay in getting them into full motion. For the next inner layer the delay is rather greater, and so on. But ultimately the motion penetrates everywhere equally, and everything is in a steady state.

But while the process of starting the wheels was

o

going on, a slip took place round the contour A B C D, and round every concentric contour inside it; the periphery of the positive wheels moving in a direction opposite to that of the wheels in contact with the rack, and so suggesting the opposite induced current excited at " make " in the substance of a conductor near a growing current, or generally in an increasing magnetic field.

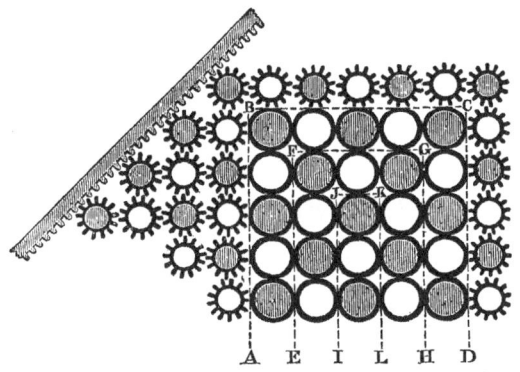

FIG. 43.—Diagram illustrating the way in which an induced current arises in a mass of metal immersed in an increasing magnetic field ; also how it decays. The dotted lines A B C D, E F G H, I J K L, are successive lines of slip.

The penetration of the motion deeper and deeper and the gradual dying away of all slip, illustrate also the mode in which this induced current arises and gradually dies away; becoming *nil* as soon as the magnetic field (*i.e.* the rotation) has penetrated to the interior of all conductors and become permanently established there as elsewhere.

Suppose the motion of the rack now stopped : all the cogged wheels stop too, though it may be with a jerk and some violence and oscillation due to their momentum ; but those inside the contour A B C D will continue moving for a little longer.   The outside layer of this region will slip in such direction as to illustrate the direct induced current at "break" and will begin to stop first ; the slip and the stop gradually penetrating inwards just as happened during the inverse process, until all trace of rotation ceases.   This inverse slipping process is the direct induced current at "break."

104. Through a perfect conductor the disturbance could never pass, for the slip of the dielectric wheels on its outer skin would be perfect, and would never penetrate any deeper.   A superficial current lasting for ever, or rather as long as the magnetic field (the rotation of the dielectric wheels) lasts, is all that would be excited, and it would be a perfect magnetic screen to any dielectric beyond and inclosed by it.   Such a perfect conductor is represented by the empty space in Fig. 38.   A magnetic field or spin excited outside that space would never reach the set of wheels inclosed and protected by it (§ 153).

105. It will now be perceived that a fly-wheel in rotation is the mechanical analogue of magnetism, or more definitely of a section of a line (or tube) of magnetic force ; and that a brake applied to such a

fly-wheel, with consequent slip, dissipation of energy, and production of heat, is in some sort a mechanical analogue of an electric current.

The field is regarded as full of geared elastic vortices or whirls, some of which are cogged together, so to speak, while others are merely pressed together by smooth rims. It is among these latter that slip is possible, and in the regions occupied by them that currents exist ; the energy dissipated here being transmitted through the non-slippery or dielectric regions from the source of power, just as energy is transmitted from a steam-engine through mill-work or shafting to the various places where it is dissipated by friction.

Transfer of Energy to a Distance.

106. Let us now try to understand the use of a telegraph wire from this point of view. Given the means of exciting a magnetic field at one place, how is one to transmit it to another place so as to move magnetic needles or make other signals there ? The first idea of a method might be that inasmuch as no perfect conductor, or absolute magnetic screen, intervenes, the field of any magnet is infinite in extent, and consequently already reaches the distant place. Have *here* a long iron bar capable of being magnetized and demagnetized at pleasure, and have *there* a very

sensitive magnetometer, and the thing is done. I see no reason why, under certain circumstances, this mode of signalling without wire over short distances should not be attempted. But an obvious objection to it is that the effect produced by a given magnet varies inversely with the cube of its distance, so that a few miles away the force of a magnet, even several yards long, is terribly weak.

The next idea might be to carry some of the magnetic lines of force to the distant place by means of an iron rod or wire. A soft iron wire transmits them so much better than air, that an arrangement of a very elongated loop of iron, with a magnetizing coil upon it at one end, and a receiving coil upon it at the other, might serve to establish some connection between the two places, and enable primary currents at this end to produce secondary induced currents at that. This would be a magnetic telegraph in which a magnetic whirl only is propagated along the wire, and a current excited by it at the distant place.

The current loop and the magnetic loop are, however, reciprocal: and the next idea might be that instead of a long magnetic loop with little current loops threading it at either end, it might be better to have a long current loop with little magnetic loops threading it at either end ; and that is exactly what we have in the electric telegraph.

It is a better plan for this reason. Iron conducts magnetism, say, a thousand times better than air, but by no means infinitely better ; hence from a long loop of iron a great many lines of force would leak, and would take a shorter circuit through air.

But copper wire conducts electricity almost infinitely better than gutta-percha or porcelain. That is why an electric telegraph is better than a magnetic. Lead or German silver conducts a million times better than dilute sulphuric acid, and yet it would be very unsatisfactory to have to signal through an ocean of dilute acid with an uncovered lead or German silver wire. The percentage of loss in the case of a corresponding magnetic circuit of iron would be far greater still.

107. But now what is it precisely that the wire of the electric telegraph does ? A magnetic field at this end is made to excite a magnetic field at the other end with very little loss ; it is nearly all concentrated upon the other end by means of the wire. Somehow the wire enables us to transmit the magnetic effect exactly in the direction we wish, and to reproduce it where we please. It is all very well to talk of a current going by the wire ; but now that we are regarding a current as a mere slipping of the gearing of the magnetic medium, we see that nothing really travels along the wire at all.

Suppose, for simplicity, the wire to be a perfect

conductor : the magnetic gearing, which penetrates everywhere else and communicates the magnetic spin, breaks down at its surface and nothing is transmitted into the wire. The wire, in fact, is electrically nothing but a line of slip of arbitrary shape penetrating and modifying the magnetic field. Up to its surface magnetic propagation goes on : at its surface it stops.

Well, how does that help us ? How can this fact enable the wire to transmit the signals ? That is what we have got to see.

Refer back to Fig. 30, B, and look at it in the light of Fig. 39, taking the latter as a section only, so that each of the little arrows in Fig. 30, B, represents the same fact as the cog-wheels do in Fig. 39. Further, imagine the rack in Fig. 39 to be removed, or replaced by a perfectly smooth rod, and let the wheels rotate just as they would have done had the rack been pushed along; then on the surface of the stationary or non-existent rod we have the state of slip which we have now learned to recognize as current. What now is the function of the rod or of the space it occupies ? It permits free rotation of the wheel-work on either side of it in opposite directions ; whereas if the rod or space were removed, and the cogs allowed to gear across, they would at once jamb, and the action would be stopped.

Take away the long conductor, and the only

magnetic field you can have is the ordinary state
of spin round lines of force, rapidly becoming fainter
and fainter with distance ; but with the long conductor, acting as core to the whirls, you have a
different state of things altogether. Away from the
wire the field is weak, but close to it it is strong : all
round the wire is an intense magnetic field as in
Fig. 30, B, and this state of things extends the whole
length of the wire undiminished by distance.

In order that a wire may act in this manner, it
must form a closed circuit, and it must have a propelling arrangement at some part of it able to excite
the vortex cores upon it just there. Given these
conditions, nothing can stop the vortex cores from
travelling right along the wire, however long it may
be, and producing their effect at the distant station.

It is not very easy to draw a diagram of the
arrangement, because it entails such a number of
wheels, and because the diminution of spin with
distance is not well represented by them. But the
diagram may be imagined thus :—

Let the rack in Fig. 39 be regarded as an infinitesimal portion of a long circuit extending to
New York and back. It may be considered smooth
or cogged as one pleases : to avoid the idea of any
material transfer it is better to think of it as smooth.
At some one point, by means of a battery, or a
dynamo, or any other electromotive arrangement,

excite in a few of the wheels the motion which in Fig. 39 would occur if the rack were pushed. Out and away through the dielectric the motion transmitted by the cog-wheels spreads, at a pace which for the present we may consider infinite, but which we shall learn later is the velocity of light. At a distance from the wire the spin is small—it diminishes as the inverse distance ; but close to the wire, all the way along, the opposing cogs are kept apart, and there the spin is most intense. Right along the wire flashes the vortex-like motion ; all the way along it is surrounded with rings of whirl, as in Fig. 30, B, and by concentrating some quantity of this whirl into small compass at the distant station, a visible motion or signal is produced.

This is the function of the wire, it guides the effect transmitted through the dielectric. The wire transmits nothing—it is the insulating sheath that transmits all the energy : the wire directs it on its way by holding asunder the mutually opposing gearing of the dielectric.

So much if the wire be perfectly conducting. If it be an ordinary wire it acts in just the same way, only that the slip on its surface is not perfect—it is accompanied, as it were, by friction ; and so its own wheel-work is more or less set in motion in concentric cylinders, except just the axis of the wire, which is undisturbed. The process of transmission is just the same as in the case of a perfect conductor, except for two things :—First, the having to start the wheel-work inside the metal may delay the process a little, especially if the wheels be very massive—as they are in the case of iron. That is the first distinction—the delay caused by the conductor being either very thick or very magnetic, or both. The second distinction is that the friction and slip on the imperfect conductor dissipates some of the original supply of energy into heat, and that which is transmitted is accordingly less than with a perfect conductor. But observe still that although the wire now dissipates energy, it transmits none ; all that enters it is lost : the dielectric alone, with its cogged gearing, transmits energy to the distant station (§§ 42—45).

Later on we shall learn that the gearing in a dielectric is not rigid, but elastic, and that this is why a certain time is required for transmission—why a definite velocity of transmission exists. We may be said to have learnt it already, indeed ; and we have also learned that some dielectrics are less rigid than others gutta-percha less than air, for instance, (see §§ 16 and 23), and accordingly transmit the disturbances more slowly ; but always, as we shall find, with the appropriate velocity of light, so far as the dielectric alone is concerned (§§ 133 *et seq.*).

Mechanical Force acting on a Conductor conveying a Current.

108. In Fig. 41 the conducting portion of a circuit is shown with its appropriate opposite rotations on either side of it. Now superpose a uniform rotation all in one direction upon this, so as to increase the spin on one side of the conductor and diminish it on the other; in other words, immerse the circuit in a

FIG. 44.—Two parallel conductors conveying equal currents in one direction and getting pushed together by the centrifugal force of the outside whirls, no whirl existing between them. The length of the arrows again suggests the distribution of magnetism in the conductors. Fig. 40 showed the correlative repulsion of opposite currents.

magnetic field.   Immediately the extra centrifugal force on one side will urge any movable part of the conductor from the stronger to the weaker portion of the field.   And whether there be any movable portion or not, the whole circuit will *tend* to expand if the

superposed magnetic whirl agree in direction with
the whirl already inside ; while it will tend to
contract if the superposed whirl agree with that
outside.

The field for a direct and return circuit may be
similarly drawn by superposition of their separate
whirls (see Fig. 40). In this case there is strong
centrifugal force of the whirl between the wires, while
outside there is next to no whirl at all. Hence the
wires tend to get driven apart ; and so it becomes
evident why a circuit tends to expand so as to inclose
the largest possible area, even if no other magnetic
field than its own be acting on it. The circuit shown
in Fig. 41, for instance, tends to expand even without
any superposed magnetic field, simply because the
whirl inside is more concentrated, and therefore more
intense, than the whirl outside.

Also if two circuits are arranged near each other
in a plane, with their currents in opposite directions,
they will more or less neutralize each other's effect on
the space between them, causing (if equal) a region of
no spin there. The two conductors will thus get
urged together by the unbalanced pressure of the
centrifugal force due to the whirl on the other side ;
or, currents in the same direction attract.

109. As for the effect of iron introduced into a
circuit, it brings into the region of space it occupies
some hundred or thousand times as many lines of

whirl as were there before, and these naturally contribute mightily to the effects, both those exhibiting mechanical force and those exhibiting inertia.

When one says, as roughly one may do, that iron brings 1000 fresh lines into the field, one means that, for every whirl otherwise excited, 1000 more are faced round in the iron. And this process goes on while the field is increasing in strength until the total number of whirls in the iron begins to be called upon ; when this point is reached the rate of addition is not maintained, and the iron is said to show signs of saturation. Ultimately, if ever all its whirls were faced round, the iron would be quite saturated ; but long before this point is reached another cause is likely to make itself felt, viz. the falling off in the strength of the whirls already faced round, by the action of the strong magnetic induction, which is all the time acting so as to weaken the iron currents so far as it is able. And thus at a certain point hitherto unreached by experiment the iron may not only fail to increase the strength of the field any more, but may actually begin to diminish it. That is to say, its permeability may conceivably become less than 1, as if it were a diamagnetic substance (cf. § 81).

The easiest way to picture the effect of iron is to think of its wheels as some hundred or thousand times as massive as those of air, so that their energy and momentum are very great (cf. § 97).

That which is commonly called magnetic permeability, and denoted by μ (§ 82), may in fact be thought of as a kind of inertia, an inertia per unit volume ; in other words, a *density*—an ethereal density ; though how it comes to pass that the ether inside iron is endowed with so great inertia one cannot say. Perhaps it is that the iron atoms themselves revolve with the electricity (§ 94), perhaps it is something quite different. Whatever the peculiar behaviour of iron, nickel, &c., be due to, it must be something profoundly interesting and important as soon as our knowledge of their molecular structure enables us to perceive its nature.

Induction in Conductors not originally carrying Currents but moving in a Magnetic Field.

110. To explain the currents induced in a conductor moving through a uniform magnetic field is not quite easy, because none of the diagrams lend themselves naturally and simply to the idea of circuits changing in form or size.

If we take a rigid circuit in a magnetic field, like Fig. 45, and revolve it out of its plane 180°, it is obvious that a current will be excited in it, for the process is essentially the same as if the conductor were kept still and the field reversed.

But to understand the current excited in a closed

circuit when a portion of it moves across the lines so
as to embrace a greater number of them, one has to
take into account the fact that the inside whirls are
expanding and doing work in forcing the conductor
away, while the outer whirls are resisting the motion,
and being thereby compressed and rendered more energetic. Thus the wheels inside revolve slightly slower

FIG. 45.—Section of a uniform magnetic field with two rails and a slider in it. If
the slider be moved to or fro, the wheels inside get initially compressed or
extended, and thereby gain or lose energy respectively, thus exciting the state of
slip known as induced current.

as the circuit expands, and those on the other side
the slider revolve slightly quicker. Both these processes cause a slipping of the gearing, first all round
the inside, and then all through the substance of the
wire, whereby positive electricity moves forward in
one direction round the circuit, the negative moving
oppositely ; and so a current is accounted for. It is

not to be supposed, however, that any finite expansion
of the wheels really occurs : the motion is rapidly
equalized by diffusion through the wire, and fresh
wheels come in round it from outside ; hence directly
after the conductor has stopped moving the field is
again steady, but with many more wheels inside the
contour than it possessed at first.

*Representation of an Electrostatic Field again, and
superposition of it on a perpendicular Magnetic Field.*

111. An electrostatic strain is, we know, caused by
a displacement of positive electricity one way along
the lines of force, and by an equal displacement of
negative the other way.   The process was indicated
crudely in Fig. 7A ; we may now represent it rather
more fully with the help of our elastic cells by
Fig. 46.

Here the positive cells have been pulled one way, the
negative the other way ; and when the distorting force
is removed, the medium tends to spring back to its
normal condition, exerting an obvious tension on
bodies attached to it in the direction of its lines of force,
its elongated direction, and an obvious pressure in all
perpendicular directions, its compressed directions.

Now, if all the cells are full of parallel whirls, as in
the preceding magnetic diagrams, it is not improbable

that this electrostatic distortion or "shear" of the medium may affect its magnetic properties slightly, and that, if the direction of electrostatic strain were

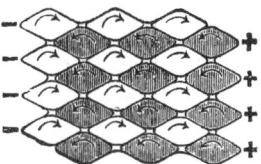

FIG. 46.—A portion of an electrostatic field between two oppositely charged bodies, with its lines of force going from right to left, and showing a tension along and a pressure at right angles to them, due to the elasticity of the cells (which elasticity may be due to their containing fluid in a state of whirl, see § 156). Magnetic lines of force perpendicular to the paper are also shown in section. While this magnetic field was being excited and propagated from below upwards, a slight strain would be produced in the elastic cells, like but immensely less than that shown ; as contrasted with its normal condition (Fig. 37). Conversely, while this electrostatic strain was being produced, the positive whirls would be infinitesimally quickened and the negative ones retarded during the displacement, thus producing a minute magnetic effect. If the medium is not magnetized, the whirls are not necessarily absent, only faced all ways.

rapidly reversed, a small magnetic oscillation would also ensue ; but the exact details of these mutual actions are difficult to specify at present.

Disruptive Discharge.

112. Disruptive discharge may be thought of as a pulling of the shaded cells violently along past the others ; the process being accompanied by a true disruption—a sort of electrolysis—of the medium, and a passage of the two electricities in opposite directions along the line of discharge.

P

Consider the locomotion of any one horizontal row of shaded cells in Fig. 46 during the occurrence of such a disruption of the medium. The cells slide on towards the right, and, as they slide, the spin of the negative cells above them is retarded while that of those below them is accelerated ; consequently a true magnetic effect is produced, just like that accompanying a current, and a disruptive discharge has therefore all the magnetic properties of a current.

Effects of a Moving Charge.

113. This locomotion of a set of positive cells, or of negative cells the other way, as just considered, is very near akin to the motion of a charge through a dielectric medium.

When a charged body moves along with extreme rapidity, it can be thought of as exciting a rotation in the cells most closely in contact with it, greater than that which it excites in the opposite kind of cells, and thus produces the whirl proper to a magnetic field. Thus does a moving charge behave just like a current of a certain strength.

It may be, indeed, that this is the customary way of exciting a voltaic current ; for the chemical forces in a cell cause a locomotion of charged atoms, and thus set up a field, which, spreading out in the way Prof. Poynting has sketched, (§ 42) reaches every part of the metallic circuit and excites the current there.

Electrostatic Effects of a Moving or Varying Magnetic Field.

114. Just as we have seen that a moving or varying electrostatic field may produce slight magnetic effects, so one can perceive that a moving or varying magnetic field brings about something of the nature of an electrostatic strain.

For a spreading out field is continually propagating the rotation on from one layer of wheels to the next. It there is any slip, we thus get induced currents (Fig. 43), and the rate of propagation is comparatively slow, being a kind of diffusion; but even if there is not any slip, yet, unless the wheel-work is absolutely rigid, the rate of propagation will not be infinite. The actual rate of propagation is very great, which shows that the rigidity or elasticity of the wheels is very high in proportion to their inertia, but it is not infinite; and accordingly the propagation of rotation is accompanied by a temporary strain. One part of the field is in full spin, another more distant part is as yet unreached by the spin; between the two we have the region of strain, the wheel-work being distorted a little while taking up

P 2

the motion. Thus does a spreading out magnetic field cause a slight and temporary electrostatic strain, at right angles both to the direction of the lines of force and to the direction of their advance.

Generation of a Magnetic Field. Induction in closed Circuits.

115. Picture to one's self an unmagnetized piece of iron ; its whirls are all existent, but they are shut up into little closed circuits, and so produce no external effect ; magnetize it slightly, and some of the closed circuits open out and expand, with one portion of them in the air. Magnetize it strongly, and we have a whole set of them opened out into vortex cores, still with the whirl round them, and constituting the common magnetic lines of force. There is no need to think of iron and steel in this connection. In air or any substance the whirls are still present, though much fewer or feebler, and their axes ordinarily form little closed circuits—it may be inside the atoms themselves. But wrap a current-conveying wire round them, and at once they open out into the lines of force proper to a circular current.

Again, think of an iron ring, or a hank of wire as bought at an ironmonger's : wrap a copper wire several times round it, as a segment of a Gramme ring is wound (Fig. 47) and pass a current. The closed vortices in the iron at once expand : a portion of each flashes out and across the air-space inclosed by the ring (not by any means confining itself to a plane, of course), and enters the ring on the opposite side ; so that directly the current is steady the lines all lie inside the iron again, but now inclosing an area—the area of the ring—instead of being shut up into infinitesimal links. In a sense the iron is still unmagnetized, for its lines of force still form closed contours within it, and none protrude any part of themselves into the air,

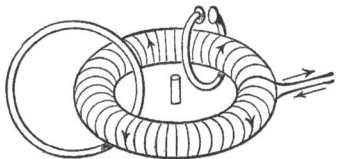

FIG. 47.—Closed magnetic circuit like Fig. 42, with a single-ring secondary circuit, and another open secondary loop ; also with a short conducting-rod standing up in it.

except for irregularities. But in another sense it is highly and permanently magnetized round and round in itself, the magnetism being not easy to get out of it again, except by judiciously arranged reverse currents.

It is now like one great electric vortex ring instead of like a confused jumble of microscopic ones. Its section was shown in Fig. 42. (See also Appendix (*d*) and (*n*).)

During the variable period, while the current is increasing in strength, or while it is being reversed, the region inclosed by the ring and all around it is full of myriads of expanding lines of force flashing across, broadside on, from one side of the iron to the other, and there stopping. It is the presence of these moving lines, changing rapidly from a " simply-connected " into a " multiply-connected " state, or *vice versâ*, which causes the powerful induced currents of " secondary generators."

In every case of varying magnetic field, in fact, we have lines moving broadside on, propagating their whirl, and more or less disturbing the medium through which they move.

Next consider a moving or spinning magnet. Its lines travel with it, and, being closed curves, they also must move broadside through the field, so that in this case we may expect just the same effect as can be obtained from a varying magnetic field.

If a broadside-moving line of force cut across a conductor, its motion is delayed, for its wheels slip and only gradually get up a whirl inside the ill-geared substance ; thus, as we know, causing an induced current (see § 103).

If a conducting ring is looped with the iron ring previously mentioned, as a snap-hook is looped with an eye, then every expanding vortex, while the ring is being magnetized, has necessarily to cut through the conducting ring once and no more, no matter what its shape or size. The electromotive force of induction is in this case therefore perfectly definite, and simply proportional to the number of turns made by the secondary round the core of the ring (Fig. 47). Instead of supposing a closed conducting secondary circuit, imagine an open one : there is still an E.M.F. in it, though rather less than before because a few of the expanding lines flash through the gap and produce no effect; so electricity must surge to and fro in the conductor, while the ring is being magnetized and demagnetized, as water surges up and down in a tilted trough, and a small condenser attached to the free ends will be alternately charged and discharged. The gap might become so large that nothing is left but a short rod (Fig. 47): in this also similar oscillations would occur.

But now suppose no secondary conductor at all ; nothing but dielectric inclosed by the ring. In it there must be an electric displacement excited every time the magnetism of the ring is reversed. It may be an oscillatory displacement, but still on the whole in one direction during rise of magnetism, and in an opposite direction during reversal of magnetism. A charged body delicately suspended within the ring may feel the effect of the minute electrostatic strain so magnetically produced.[1]

116. To see the *mode* in which an electrostatic displacement arises in the space embraced by the ring

[1] See *Phil. Mag.* June, 1889.

we have only to turn to Fig. 42, and look at the set of wheels along the line A B separating one half the section from the other. They cannot steadily rotate either way, for they are urged in opposite directions by the two halves ; in other words, there is no magnetic field anywhere near such a ring, as is well known ; but, nevertheless, during a change of magnetism, while the whirls inside are changing in speed, the rub on the dielectric necessary for checking the outer wheels of the conductor is either increased or diminished ; and if the wheels have any elastic "give" in them, as we know they have, (§ 111) the electrostatic strain in the field is thereby altered during the varying stage of the magnetism.

PART IV.

RADIATION.

CHAPTER XII.

RELATION OF ETHER TO ELECTRICITY.

117. So far as we have been able to understand and explain electrical phenomena, it has been by assuming the existence of a medium endowed with certain mechanical or *quasi*-mechanical properties, such as mobility (§ 12), incompressibility or infinite elasticity of volume (§ 5), combined with a certain amount of plasticity or finite elasticity of shape (§ 9). We also imagined the medium as composed of two opposite constituents, which we called positive and negative electricity respectively (§ 90), and which were connected in such a way that whatever one did the other tended to do the precise opposite. Further, we were led to endow each of these constituents with a certain amount of inertia (§§ 38 and 88), and we recognized something of the nature of friction between each constituent and ordinary matter (§§ 28 and 63).

Broadly speaking we may say—

(1) That *friction* makes itself conspicuous in the

discussion of current-electricity or the properties of conductors, and that the laws of it are summarized in the statement known by the name of Ohm, viz. that the current through a given conductor is proportional to the force that drives it, or that the opposition force exerted by a conductor upon a current is simply proportional to the strength of that current.

(2) That *elasticity* is recognized as necessary when studying the facts of electrostatics or the properties of insulators—electric displacement and recoil, or charge and discharge : the laws having been studied by Faraday, and the relative pliability (or shearability if there were such a word) of the medium in different substances being measured and stated in terms of that of air as their specific inductive capacity, K.

(3) That *inertia* is brought into prominence by the facts of magnetism, studied chiefly perhaps by Thomson, who has called the relative density of the medium in different substances their magnetic permeability or magnetic inductive capacity ; the ratio of its value for any substance to its value for common air being called μ.

(4) That the *doubleness of constitution* of the medium—its being composed of two precisely opposite entities—is suggested by the facts of electrolysis, by the absence of mechanical momentum in currents and magnets, and by the difficulty of otherwise conceiving a medium endowed with rigidity

which yet is perfectly fluid to masses of matter moving through it.

118. With the hypothesis of doubleness of constitution this last-mentioned difficulty disappears. The ether as a whole may be perfectly fluid and allow bodies to pass through it without resistance, while its two components may be elastically attached together and may resist any forces tending to separate them with any required rigidity. It is like the difference between passing one's hand through water, and chemically decomposing it ; it is like the difference between waving a piece of canvas about, and tearing it into its constituent threads.

To put the matter boldly and baldly : we are familiar with the conceptions of matter and of ether, and it is known that the two things react on each other in some way, so that although a free portion of the ether appears to move freely through matter, yet another portion appears to move with matter as if bound to it. This mode of regarding the facts is as old as Fresnel (§ 184). We now proceed a step further, and analyze the ether into two constituents—two equal opposite constituents—each endowed with inertia, and each connected to the other by elastic ties : ties which the presence of gross matter in general weakens and in some cases dissolves. The two constituents are called positive and negative electricity respectively ; and of these two electricities we imagine the ether to be composed. The tie between them is dissolved in metals, it is relaxed or made less rigid in ordinary insulators. The specific inductive capacity of a substance means the reciprocal of the rigidity of its doubly constituted ether. Let us call this rigidity k, so that $k = \dfrac{1}{K}$.

The neighbourhood of gross matter seems also to render ether more *dense*. It is difficult to suppose that it can really condense an incompressible fluid, but it may load it or otherwise modify it so as to produce the *effect* of increased density. In iron this density reaches its highest known value, and in all substances the density or inertia per unit volume of their ether may be denoted by μ, and called their magnetic permeability.[1]

119. Let it be understood what we are doing. In Part I. we discussed effects very analogous to those which would be produced by an elastic incompressible medium (roughly like india-rubber or jelly); that is, we were led to postulate a medium possessing elasticity or something very analogous to elasticity. In Parts II. and III. we discussed effects suggesting, and more or less necessitating, the idea of a property of the medium very analogous to inertia ; and we were also led to postulate a doubleness of constitution for the medium, so that shearing strains may go on in it and yet it be perfectly fluid as a whole. We are

---

[1] Strictly speaking, the density is more likely to be $4\pi\mu$, and the rigidity $4\pi/k$; but the 4π is omitted for simplicity.

now pushing these analogies and ideas into greater
definiteness and baldness of statement. We already
know of a continuous incompressible fluid filling all
space, and we call it the ether. Let us suppose that
it is composed of, and by electromotive force ana-
lyzable into, two constituents ; let these constituents
cling together with a certain tenacity, so that the medium shall have an electromotive elasticity, though
mechanically quite fluid ; and let each constituent possess inertia, or something so like inertia as to produce
similar effects. Making this hypothesis, electrical
effects are to a certain extent explained. Not
ultimately, indeed—few things can be explained
ultimately : not even as ultimately as could be
wished ; for the nature of the connection between the
two constituents of the ether and between the ether
and gross matter—the nature of the force, that is, and
the nature of the inertia—remains untouched. This is
a limitation to be clearly admitted ; but if that were
the only one—if all else in the hypothesis were true—
we should do well, and a distinct step would have been
gained. It is hardly to be hoped that this is so—
hardly to be expected that the bald statement above
is more than a kind of parody of the truth ; nevertheless, supposing it only a parody, supposing what we
call electromotive elasticity and inertia are things
capable of clearer conception and more adequate statement, (§ 156) yet, inasmuch as they correspond to and

represent a real analogy, and inasmuch as we find that a medium so constructed would behave in a very electrical manner, and might in conjunction with matter be capable of giving rise to all known electrical phenomena, we are bound to follow out the conception into other regions, and see whether any other abstruse phenomena, not commonly recognized as electrical, will not also fall into the dominion of this hypothetical substance and be equally explained by it (§ 8). This is what we shall now proceed to do.

120. Before beginning, however, let me just say what I mean by "electromotive elasticity." It might be called chemical elasticity, or molecular elasticity. There is a well-known distinction between electro-motive force and ordinary matter-moving force. The one acts upon electricity, straining or moving or, in general, "displacing" it ; the other acts upon matter, displacing it. The nature of neither force can be considered known, but crudely we may say that as electricity is to matter so is electromotive force to common mechanical force ; so also is electromotive elasticity to the common shape-elasticity or rigidity of ordinary matter : so perhaps, once more, may electrical inertia be to ordinary inertia.

Inertia is defined as the ratio of force to acceleration ; similarly electric inertia is the ratio of electromotive force to the acceleration of electric

displacement. It is quite possible that electric inertia and ordinary inertia are the same thing, just as
electric energy is the same with mechanical energy. If
this were known to be so, it would be a step upward
towards a mechanical explanation ; but it is by no
means necessarily or certainly so ; and whether it be
so or not, the analogy undoubtedly holds, and may be
fruitfully pursued.

And as to " electromotive elasticity," one may say
that pure water or gas is electromotively elastic, though
mechanically limpid ; each resists electric forces up to
a certain limit of tenacity, beyond which it is broken ;
and it recoils when they are withdrawn. Glass acts
in the same way, but that happens to be mechanically
elastic too. Its mechanical elasticity and tenacity
have, however, nothing to do with its electric elasticity
and tenacity.

One perceives in a general way why fluids can be
electrically, or chemically, or molecularly elastic : it is
because their molecules are doubly or multiply composed, and the constituent atoms cling together, while
the several molecules are free of one another. Mechanical forces deal with the molecule as a whole, and to
them the substance is fluid ; electrical or chemical
forces deal with the constituents of the molecule,
setting up between them a shearing strain and endeavouring to tear them asunder. To such forces,
therefore, the fluid is elastic and tenacious up to a

Q

certain limit. Extend this view of things to the constitution of the ether, and one has at least a definite position whence to further proceed.

121. It may be convenient and not impertinent here to say that a student might find it a help to re-read Parts I. and II. in the light of what has just been said : remembering that, for the sake of simplicity, only the simple fact of an elastic medium was at first contemplated and insisted on ; no attempt being made to devise a mechanism for its elasticity by considering it as composed of two constituents. Hence the manifest artificiality of such figures as Fig. 6, where fixed beams are introduced to serve as the support of the elastic connections. But it is pretty obvious now, and it has been already indicated in Fig. 7A, that a closer analogy will be obtained by considering two sets of beads arranged in alternate parallel rows connected by elastic threads, and displaced simultaneously in opposite directions. A still further progressive analogy is attempted in Fig. 46. We have gradually passed, therefore, from a sort of one-fluid theory to a modified two-fluid theory ; believing it to be in some sense or other nearer the truth.

Recovery of the Medium from Strain.

122. We have now to consider the behaviour of a medium endowed with an elastic rigidity, k, and a density, μ, subject to displacements or strains. One

obvious fact is that when the distorting force is removed the medium will spring back to its old position, overshoot it on the other side, spring back again, and thus continue oscillating till the original energy is rubbed away by viscosity or internal friction. If the viscosity is very considerable, it will not be able so to oscillate ; it will then merely slide back in a dead-beat manner towards its unstrained state, taking a theoretically infinite time to get completely back, and practically restoring itself to something very near its original state in what may be quite a short time. The recovery may in fact be either a brisk recoil or a leak of any degree of slowness, according to the amount of viscosity as compared with the inertia and elasticity (§ 19).

The matter is one of simple mechanics. It is a case of simple harmonic motion modified by a friction proportional to the speed. The electrical case is simpler than any mechanical one, for two reasons : first, because so long as capacity is constant (and no variation has yet been discovered) Hooke's law will be accurately obeyed—restoring force will be accurately proportional to displacement ; secondly, because for all conductors which obey Ohm's law (and no true conductor is known to disobey it) the friction force is accurately proportional to the first power of velocity.

123. There are two, or perhaps one may say three,

main cases. First, where the friction is great. In
that case the recovery is of the nature of a slow leak,
according to a decreasing geometrical progression
or a logarithmic curve ; the logarithmic decrement
being independent of the inertia, and being equal
to the quotient of the elasticity and the resistance
coefficients.

As the resistance is made less, the recovery becomes quicker and quicker until inertia begins to
prominently assert its effect and to once more
lengthen out the time of final recovery by carrying
the recoiling matter beyond its natural position, and
so prolonging the disturbance by oscillations. The
quickest recovery possible is obtained just before
these oscillations begin ; and it can be shown that
this is when the resistance coefficient is equal to twice
the geometric mean of the elasticity and the inertia.
One may consider this to be the second main case.
The recoil is then exactly dead-beat, and occurs in the
minimum of time.

The third principal case is when the resistance is
quite small, and when the recovery is therefore
distinctly oscillatory. If the viscosity were really
zero, the motion would be simply harmonic for ever,
unless . some other mode of dissipating energy were
provided ; but if some such mode were provided, or
if the viscosity had a finite value, then the vibrations would be simply harmonic with a dying out

amplitude, the extremities of all the swings lying on a logarithmic curve. In such a case as this, the rate of swing is practically independent of friction; it depends only on elasticity and inertia; and, as is well known for simple harmonic motion, the time of a complete swing is 2π times the square root of the ratio of inertia and elasticity coefficients.

124. Making the statement more electrically concrete, we may consider a circuit with a certain amount of stored-up potential energy or electrical strain in it: for instance, a charged Leyden jar provided with a nearly complete discharge circuit. The main elastic coefficient here is the reciprocal of the capacity of the jar: the more capacious the jar the more "pliable" it is—the less force of recoil for a given displacement,— so that capacity is the inverse of rigidity. The main inertia coefficient is that which is known electrically as the "self-induction" of the circuit: it involves the inertia of all the displaced matter and ether, of everything which will be moved or disturbed when the jar is discharged. It is not a very simple thing to calculate its value in any given case; still it can be done, and the general idea is plain enough without understanding the exact function and importance of every portion of the surrounding space. (See Appendix.)

Corresponding, then, to the well-known simple harmonic $T = 2\pi \sqrt{\frac{m}{k}}$, we have, writing L for the self-induction or inertia of the circuit, and S for its capacity or inverse elasticity constant,

$$T = 2\pi \sqrt{LS}.$$

This, therefore, is the time of a complete swing. Directly the jar is discharged, these oscillations begin, and they continue like the vibration of a tuning-fork until they are damped out of existence by viscosity and other modes of dissipation of energy.

125. But now just consider a tuning-fork. Suppose its substance were absolutely unviscous, would it go on vibrating for ever ? In a vacuum it might : in air it certainly would not. And why not ? Because it is surrounded by a medium capable of taking up vibrations and of propagating them outwards without limit. The existence of a vibrating body in a suitable medium means the carving of that medium into a succession of waves and the transmission of these waves away into space or into absorbing obstacles. It means, therefore, the conveyance away of the energy of the vibrating body, and its subsequent appearance in some other form wherever the radiating waves are quenched (§ 141).

The laws of this kind of wave-propagation are well known ; the rate at which waves travel through the medium depends not at all on any properties of the original vibrating body, the source of the disturbance ; it depends solely on the properties of the medium.

They travel at a rate precisely equal to the square root of the ratio of its elasticity to its density.

Although the speed of travel is thus fixed independently of the source, the length of the individual waves is not so independent. The length of the waves depends both on the rate at which they travel and on the rate at which the source vibrates. It is well known and immediately obvious that the length of each wave is simply equal to the product of the speed of travel into the time of one vibration.

126. But not every medium is able to convey every kind of vibration. It may be that the mode of vibration of a body is entirely other than that which the medium surrounding it can convey : in that case no dissipation of energy by wave-propagation can result, no radiation will be excited. The only kind of radiation which common fluids are mechanically able to transmit is well known : it is that which appeals to our ears as sound. The elasticity concerned in such disturbance as this is mere volume elasticity or incompressibility. But electrical experiments (the Cavendish experiment, § 4, and Faraday's ice-pail experiment) prove the ether to be enormously—perhaps absolutely—incompressible ; and if so, such vibrations as these would travel with infinite speed and not carve proper waves at all.

Conceivably (I should like to say probably) *gravitation* is transmitted by such longitudinal impulses or thrusts, and in that case it is nearly or quite instantaneous ; and the rate at which it travels, if finite, can be determined by a still more accurate repetition of the Cavendish experiment than has yet been made ; but true radiation transmitted by the ether cannot be of this longitudinal character. The elasticity possessed by the ether is of the nature of rigidity : it has to do with shears and distortions ; not mechanical stresses, indeed—to them it is quite limpid and resistless—but electromotive stresses : it has an electrical rigidity, and it is this which must be used in the transmission of wave-motion.

But the oscillatory discharge of a Leyden jar is precisely competent to apply to the ether these electromotive vibrations : it will shake it in the mode suitable for it to transmit ; and accordingly, from a discharging circuit, waves of electrical distortion, or transverse waves, will spread in all directions at a pace depending on the properties of the medium.

Thus, then, even with a circuit of perfect conductivity the continuance of the discharge would be limited, the energy would be dissipated ; not by friction, indeed—there would in such a circuit be no direct production of heat—it would be dissipated by radiation, dissipated in the same way as a hot body cooling, in somewhat the same way as a vibrating tuning-fork mounted on its resonant box. The energy of the vibrating body would be transferred gradually to the medium, and would by this be conveyed out and away ; its final destination being a separate question, and depending on the nature and position of the material obstacles it meets with (§ 160 ; see also a lecture on Discharge of Leyden Jar, p. 367).

CHAPTER XIII.

CONSTANTS OF THE ETHER.

Velocity of Electrical Radiation.

127. THE pace at which the radiation-waves travel depends, as we have said, solely on the properties of the medium, solely on the relation between its elasticity and its density.   The elasticity considered must be of the kind concerned in the vibrations ; but the vibrations are in this case electrical, and so electrical elasticity is the pertinent kind.   This kind of elasticity is the only one the ether possesses of finite value, and its value can be measured by electrostatic experiments. Not absolutely, unfortunately : only the relative elasticity of the ether as modified by the proximity of gross substances has yet been measured : its reciprocal being called their specific inductive capacity, or dielectric constant, K.   The absolute value of the quantity K is at present unknown, and so a convention has

arisen whereby in air it is called 1. This convention is the basis of the artificial electrostatic system of units. No one supposes, or at least no one has a right to suppose, that its value is really 1. The only rational guess at its value is one by Sir William Thomson,[1] viz. $\frac{1}{842\cdot8}$. Whether known or not, the absolute value of the dielectric constant is manifestly a legitimate problem which may any year be solved.

The other thing on which the speed of radiation waves depends is the medium's density—its electric density, if so it must be distinguished. Here, again, we do not know its absolute value. Its relative or apparent amount inside different substances is measured by magnetic experiments, and called their specific magnetic capacity, or permeability, and is denoted by μ.

Being unknown, another convention has arisen, quite incompatible with the other convention just mentioned, that its value in air shall be called 1. This convention is the basis of the artificial electro-magnetic system of units—volts, ohms, amperes, farads, and the like. Both of these conventions cannot be true : no one has the least right to suppose either true. The only rational guess at ethereal free density is one by Sir William Thomson, viz. $9\cdot36 \times 10^{-19}$.

[1] *Trans. R. S. Edin.*, xxi. 60 ; see also article "Ether," in the *Encyc. Brit.* ; and page 341 of this book.

128. Very well, then ; it being clearly understood that these two great ethereal constants, k or $\frac{1}{K}$, and μ, are neither of them at present known, but are both of them quite knowable, and may at any time become known, it remains to express the speed of wave transmission in terms of them. But it is well known that this speed is simply the square root of the ratio of elasticity to density, or

$$v = \sqrt{\frac{k}{\mu}}, \text{ or } \frac{1}{\sqrt{(K\mu)}}.$$

This then is the speed with which waves leave the discharging Leyden jar circuit, or any other circuit conveying alternating or varying currents, and travel out into space.

Not knowing either k or μ, we cannot calculate this speed directly, but we can try to observe it experimentally.

129. The first and crudest way of making the attempt would be to arrange a secondary circuit near our oscillating primary circuit, and see how soon the disturbance reached it. For instance, we might take a nearly closed loop, make it face a Leyden jar circuit across a measured distance, and then look for any interval of time between the spark of the primary discharge and the induced spark of the secondary circuit ; using a revolving mirror or what we please.

But in this way we should hardly be able to detect any time at all : the propagation is too quick.

130. Since this was first written, Dr. Hertz, of Karlsruhe, has succeeded in making a measure of velocity on this very plan. He did not indeed actually measure the time which elapsed between the closing of the primary circuit and the start of the induced current in the secondary, neither did he use a Leyden jar, but he converted the advancing waves from an electrically oscillating arrangement, excited by means of an induction coil, into stationary waves, by means of reflection at a plane metallic wall. Just as waves travelling along a rope or stretched cord are converted into stationary waves, or nodes and loops, by the interference of direct and reflected pulses : reflection taking place from the fixed end of the cord ; so waves advancing from an electrostatic oscillator, or charged body connected with the terminals of an induction coil, were reflected at the wall of the room (lined with sheet zinc on purpose to make it a conductor, and therefore a good reflector, see § 164), and by interference with the direct waves converted them into stationary nodes and loops : the interval between two nodes being half a wave-length.

By now moving the secondary circuit about, between the primary and the wall, places of maximum and minimum disturbance could be found, and thus the wave-length measured. By calculating the

oscillation period of the primary circuit (or part-cir-
cuit, for it was unclosed) an indirect measure of the
velocity of propagation was arrived at. So far as
could be told it agreed with measurements made
by other means, such as those now to be described.

131. We might next make use of the principle of
the electric telegraph, viz. the propagation of a disturbance round a single circuit from any one point of
origin. Consider a large closed circuit, either conveying or not conveying a current : introduce at any one
point a sudden change—a sudden E.M.F., for instance,
or a sudden resistance if there be a current already.
Out from that point a disturbance will spread into the
ether, just as happens in air when a blow is struck or
gun-cotton fired. A regular succession of disturbances
would carve the ether into waves : a single disturbance will merely cause a pulse or shock ; but the rate
of transmission is the same in either case, and we may
watch for the reception of the pulse at a distant station.
If the station has to be very distant in order to give
an appreciable lapse of time, a speaking-tube is desirable to prevent spreading out in all directions—to
concentrate the disturbance at the desired spot. What
a speaking-tube is to sound, that is the wire of the
circuit—the telegraph wire—to ethereal pulses.

It is a curious function, this of the telegraph wire :
it does not *convey* the pulses, it directs them. They
are conveyed wholly by the ether, at a pace deter-

mined by the properties of the ether, modified as it may be by the neighbourhood of gross matter. Any disturbance which enters the wires is rapidly dissipated into heat, and gets no further : it is the insulating medium round it which transmits the pulses to the distant station.

All this was mentioned in Part III., and an attempt was made to explain the mechanism of the process, and to illustrate in an analogical way what is going on (Chap. XI.).

The point of the matter is that currents are not propelled by end-thrusts, like water in a pipe or air in a speaking-tube, but by lateral propulsion, as by a series of rotating wheels with their axes all at right angles to the wire surrounding it as a central core, and slipping with more or less friction at its surface. This is characteristic of ether modes in general : it does not convey longitudinal waves or end-thrust pulses, like sound, but it conveys transverse vibrations or lateral pulses, like light (§ 42).

132. Without recapitulating further, we can perceive, then, that the transmission of the pulse round the circuit to its most distant parts depends mainly on the medium surrounding it. The process is somewhat as follows :—Consider two long straight parallel wires, freely suspended, and at some great distance joined together. At the near end of each, start equal opposite electromotive impulses, as by suddenly applying to

them the poles of a battery ; or apply a succession of such pulses by means of an alternating machine.   Out spread the pulses into space, starting in opposite phases from the two wires, so that at a distance from the wires the opposite pulses interfere with each other, and are practically non-existent, just as but little sound is audible at a distance from the two prongs of a freely suspended tuning-fork.   But near the wires, and especially between them, the disturbance may be considerable.   The energy emitted by the source, as it reaches each wire is dissipated, and so a fresh supply of energy goes on continually arriving at the wires, always flowing in from outside, to make up the de- ficiency.   If the wires are long enough, hardly any energy may remain by the time their distant ends are reached ; but whatever there is will still be crowding in upon the wires and getting dissipated, unless by some mechanism it be diverted and utilized to effect some visible or audible or chemical change, and so to give the desired signal (§ 107).

133. Now the pace at which this transmission of energy goes on in the direction of the wires is pretty much the same as in free space.[1]   There are various cir- cumstances which can retard it ; there are none which can accelerate it.   The circumstances which can retard it are, first, constriction of the medium by too great proximity of the two conducting wires, as, for instance,

---

[1] Appendix (o).

if they consisted of two flat ribbons close together with a mere film of dielectric between, or if one were a small-bore tube and the other its central axis or core. In such cases as this the general body of ether takes no part in the process, the energy has all to be transmitted by the constricted portion of dielectric, and the free propagation of ethereal pulses is interfered with : the propagation is no longer a simple true wave-propagation, it approximates more or less closely to a mere diffusion creep : rapid it may be, and yet without definite velocity, like the conduction of heat or the diffusion of a salt into water. One well-known effect of this is to merge successive disturbances into one another, so that their individuality, and consequently the distinctness of signalling, is lost.

134. Another circumstance which can modify rate of transmission of the pulses is ethereal inertia in the substance of the conducting wires : especially extra great inertia, as, for instance, if they are made of iron. For the dissipation of energy does not go on accurately at their outer surface ; it has usually to penetrate to a certain depth, and until it is dissipated the fresh influx of energy from behind does not fully occur. Now, so long as the value of μ for the substance of the wires is the same as that of air or free space, no important retardation is thus caused, unless the wires are very thick ; but directly the inertia in the substance of the wires is some hundred or thousand times

R

as big as that outside, it stands to reason that more time is required to get up the needful magnetic spin in its outer layer ; and so the propagation of pulses is more or less retarded.   At the same time this circumstance does not alter the character of the propagation, it does not change it from true wave velocity to a diffusion, it leaves its character unaltered ; and so the signals, though longer in coming, may arrive quite clear, independent, and distinct.   It is much the same, indeed, as if the density of the surrounding medium had been slightly increased.

I have several times mentioned the name of Prof. Poynting as one who has developed Maxwell's equations, and thrown great light upon the mode in which electro-magnetic energy is transmitted : in the same connection, and also still more prominently in connection with the general theory of telegraphy and of electro-magnetic waves, I must mention with due emphasis the name of Mr. Oliver Heaviside.   It is not for me to attempt to apportion credit, but the wide scope of his mathematical investigation into these difficult fields of research is remarkable.

135. These, then, are the main circumstances which affect the rate of transmission of a pulse from one part of a closed circuit to another : extra inertia or so-called magnetic susceptibility in the conducting substance, especially in its outer layers ; and undue constriction or throttling of the medium through

which the disturbance really has to go. Both these circumstances diminish rate of transmission ; and one (the last mentioned) modifies the law, and tends to obliterate individual features and to destroy distinctness.

Of course, besides these, the nature of the insulating medium will have an effect on the rate of propagation but that is obvious all along ; it is precisely the rate at which any given medium transmits pulses that we want to know, and on which we are thinking of making experiments. If we use gutta-percha (more accurately the ether inside gutta-percha) as our transmitting medium in an experiment, we are not to go and pretend that we have obtained a result for air.

136. The circumstances we have considered as modifying the rate of transmission are both of them adventitious circumstances, independent of the nature of the medium, and they are entirely at our own disposal. If we like to throttle our medium, or to use thick iron wires, we can do so, but there is no compulsion : and if we wish to make the experiment in the simplest manner, we shall do no such thing. We shall use thin copper wires (the thinner the better), arranged parallel to one another a fair distance apart, and we shall then observe the time which an electromotive impulse communicated at one end takes to travel to the other. Instead of using two wires, we may if we like use what comes to much the same thing, viz. a

R 2

single wire suspended at a reasonable height above the ground, as in a common land telegraph. Such a case as this is much the same as if two wires were used at a distance apart equal to about twice the height above the .ground.

The experiment, if it could be accurately made, would result in the observation of a speed of propagation equal to 3×10^{10} centimetres (300,000 kilometres, or about 185,000 miles) per second. The actual speed in practice may be less than this, by reason of the various circumstances mentioned, but it can never be greater. This, then, is the rate of transmission of transverse impulses, and therefore of transverse waves, through ether as free as it can be easily obtained.

137. The writer has succeeded in making a rough preliminary determination on this very plan, but avoiding the necessity for excessive lengths of wire by using the principle of reflection and interference to obtain stationary waves in a pair of parallel wires of known length attached as lateral appendages to a Leyden jar circuit at every discharge. Alternating pulses travel along these wires, and are reflected at their far ends, just as pulses travel along a string attached to a tuning-fork in Melde's experiment. Reflection of the pulses at the free ends of the wires is not accomplished without a considerable recoil or kick, which can be observed by the brightness of the

brush or the length of spark it gives. The length of the wires or the size of the circuit is adjusted until this recoil kick is a maximum, and the length of each wire is then taken as half a wave-length. Knowing the rate of oscillation proper to the particular Leyden jar circuit employed, a determination of the velocity of the pulses can at once be made. It agrees with what is said above.[1]

138. There are many methods known to physicists by which an indirect experimental determination of this velocity can be made. These methods have been more largely practised than the one described, but they do not determine directly the speed with which electrical pulses or waves travel : they directly determine the ratio k/μ, or, what is the same thing the product $K\mu$, and it is left to theory to say that this is really the velocity of electrical pulses in free ether. It is unnecessary to say more about them here. They are generally referred to as methods of determining the ratio " v," or the number of electrostatic units of quantity in an electro-magnetic unit ; which is a roundabout and forced mode of expression, but it serves.

[1] *Phil. Mag.* August, 1888. See also Appendix (*o*).

CHAPTER XIV.

ELECTRICAL RADIATION, OR LIGHT.

139. HAVING now described one or two possible methods of measuring the velocity of electric wave propagation, and therefore of at least the *ratio* of the two ethereal constants k and μ (or, what is the same thing, the product of the two constants K and μ) ; return to the consideration of the ordinary small discharging Leyden jar or other alternating current circuit of a moderate size, it may be a few yards or a foot or an inch in diameter.

If the alternating currents are produced artificially by some form of alternating machine, their frequency is, of course, arbitrary; but if they be automatically caused by the recoil of a given Leyden jar in a given circuit, their frequency is, as we have already said (§ 124),

$$\frac{1}{2\pi\sqrt{(LS)}} \text{ per second} ;$$

where L is the electrical inertia or self-induction of

the circuit, and where S is the capacity or reciprocal of the elasticity-constant of the jar.

140. It is not convenient here to go into the determination of the quantity L, but roughly one may say that for an ordinary open single-loop circuit it is a quantity somewhat comparable with ten or twelve times its circumference multiplied by the constant μ.[1]

The value of S has to do with the area and thickness of the condenser, being, as is well known, $\dfrac{A}{4\pi z}$ multiplied by the constant K.

The product LS in the above expression contains therefore two factors, each of linear dimensions, expressing the sizes of circuit and jar; and likewise contains a factor μK expressing the properties of the surrounding medium. Hence, so far as the ether is concerned, the above expression for frequency of vibration demands only a knowledge of the *product* of its two constants K and μ; and since this is known by the previous velocity experiments, it is easy to calculate the rate of oscillation of any given condenser-discharge. It is also easy to calculate the wave-length; for if there are n waves produced per second, and each travels with the velocity v, the length of each wave is $\dfrac{v}{n}$.

[1] See Appendix (e).

Hence the wave-length is $2\pi \sqrt{\left(\dfrac{L}{\mu} \cdot \dfrac{S}{K}\right)}$.

141. Now, if we go through these numerical calculations for an ordinary Leyden jar and discharger, we shall find waves something like, say, 50 or 100 yards long. They may plainly be of any length, according to the size of the jar and the size of the circuit. The bigger both these are the longer will be the waves.

A condenser of 1 microfarad capacity, discharging through a coil of self-induction 1 secohm, will give rise to ether waves 1900 kilometres or 1200 miles long ; and the rate of its oscillation is 157 complete swings per second.

A common pint Leyden jar discharging through a pair of tongs may start a system of ether waves each not longer than about 15 or 20 metres ; and its rate of oscillation will be something like ten million per second.

A tiny thimble-sized jar overflowing its edge may propagate waves only about 2 or 3 feet long. (See also § 157 and Appendix (k).)

142. The oscillations of current thus recognized as setting up waves have only a small duration, unless there is some means of maintaining them. How long they will last depends partly upon the conductivity of the circuit; but even in a circuit of infinite conductivity they must die out if left to themselves, from the mere fact that they dissipate their energy by radiation. One may get 10 or 100, or perhaps even 1,000,

perceptible oscillations of gradually decreasing amplitude, but the rate of oscillation is so great that their
whole duration may still be an extremely small
fraction of a second. For instance, to produce
ether waves a metre in length requires 300,000,000
oscillations per second.

To keep up continuous radiation naturally requires
a supply of energy, and unless it is so supplied the
radiation rapidly ceases. Commercial alternating
machines are artificial and cumbrous contrivances for
maintaining electrical vibrations in circuits of finite
resistance, and in despite of loss by radiation.

In most commercial circuits the loss by radiation
is probably so small a fraction of the whole dissipation
of energy as to be practically negligible ; but one is,
of course, not limited to the consideration of commercial circuits or to alternating machines as at present
invented and used. It may be possible to devise
some less direct method—some chemical method,
perhaps—for supplying energy to an oscillating circuit,
and so converting what would be a mere discharge or
flash into a continuous source of radiation.

143. So far we have only considered ordinary practicable electrical circuits, and have found their waves in
all cases pretty long, but getting distinctly shorter the
smaller we take the circuit. Continue the process of
reduction in size further, and ask what sized circuit
will give waves 6000 tenth-metres (three-fifths of a

micron, or 25 millionths of an inch) long. We have only to put $2\pi \sqrt{\left(\dfrac{L}{\mu} \cdot \dfrac{S}{K}\right)} = 0\cdot 00006$, and we find that the necessary circuit must have a self-induction in electro-magnetic units, and a capacity in electro-static units, such that their geometric mean is 10^{-5} centimetre (one-tenth of a micron). This gives us at once something near atomic dimensions for the circuit, and suggests immediately that those short ethereal waves which are able to affect the retina, and which we are accustomed to call " light," may be really excited by electrical oscillations or surgings in circuits of atomic dimensions (§§ 157-9).

If after the vibrations are once excited there is no source of energy competent to maintain them, the light production will soon cease, and we shall have merely the temporary phenomenon of phosphorescence ; but if there is an available supply of suitable energy, the electrical vibrations may continue, and the radiation may become no longer an evanescent brightness, but a steady and permanent glow.

Velocity of Electrical Radiation compared with Velocity of Light in Free Space and in Material Substances.

144. We have thus imagined the now well-known Maxwellian theory of light, viz. that it is produced by electrical vibrations, and that its waves are electrical waves.

But what justification is there for such an hypothesis beyond the mere fact which we have here insisted on, viz. that waves in all respects like light-waves except size, *i.e.* transverse vibrations travelling at a certain pace through ether, can certainly be produced temporarily in practicable circuits by familiar and very simple means, and *could* be produced of exactly the length proper to any given kind of light if only it were feasible to deal with circuits ultra-microscopic in size? The simplest point to consider is: Does light travel at the same speed as the electrical disturbances we have been considering? We described one method of measuring how fast electrical radiation travels in free space, and there are many other methods: the result was 300,000 kilometres per second. Does light travel at the same pace?

Methods of measuring the velocity of light have long been known, and the result of those measurements in free space or air is likewise 300,000 kilometres a second. The two velocities agree in free space. Hence surely light and electrical radiation are identical.

145. But there is a further test. The speed of electrical radiation was not the same in all media: it depended on the electrical elasticity and the ethereal density of the transparent substance; in other words, it was equal to the reciprocal of the geometric mean

of its specific inductive capacity and its magnetic permeability—

$$v = \frac{1}{\sqrt{(K\mu)}}.$$

Now, although the absolute value of neither K nor μ is known, yet their values relatively to air are often measured, and are known for most substances.

Also, it is easy to compare the pace at which light goes through any substance with its velocity in free space : the operation is called finding the refractive index of a substance. The refractive index means, in fact, simply the ratio of the velocity of light in space to its velocity in the given substance. The reciprocal of the index of refraction is therefore the relative velocity of light. Calling the index of refraction n, therefore, we ought, if the electrical theory of light be true, to find that $n^2 = K\mu$; or that the index of refraction of any substance is the geometric mean of its electrostatic and magnetic specific capacities.

146. That this is precisely true for all substances cannot at present be asserted. There are some substances for which it is very satisfactorily true : there are others which are apparent exceptions. It remains to examine whether they are not only apparent but real exceptions, and, if so, to what their exceptional behaviour is due.

It must be understood what the essential point is·

It has been proved by various methods, and with greater approach to exactness as the accuracy of the methods is improved, that electrical disturbances—such as the long waves emitted by any alternating machine—travel through air or free space with exactly the same velocity as light ; in other words, that there is no recognizable difference in speed between waves several hundred miles long and waves so small that a hundred thousand of them can lie in an inch. This is true in free ether, and it is a remarkable fact. If it proves anything concerning the structure of the ether, it proves that it is continuous, homogeneous, and simple beyond any other substance ; or at least that if it does possess any structural heterogeneity, the parts of which it is composed are so nearly infinitesimal that a hundred miles and the hundred-thousandth of an inch are quantities of practically the same order of magnitude so far as they are concerned : its parts are able to treat all this variety of wave-length in the same manner.

But directly one gets to deal with ordinary gross matter we know that this is certainly not the case. Ordinary matter is composed of molecules which, though small, are far from being infinitesimal. Atoms are much smaller than light-waves, indeed, but not incomparably smaller. Hence it is natural to suppose that the ether as modified by matter will be modified in a similarly heterogeneous manner ; and will accordingly not be able to treat waves of all sizes in the same way.

The speed of all waves is retarded by entering gross matter, but we should expect the smallest waves to be retarded most. The phenomenon is well marked even within the range of such light-waves as can affect the retina : the smaller waves—those which produce the sensation of blue—are more retarded, and travel a little slower, through, say, glass or water, than the somewhat larger ones which produce the sensation of red. This phenomenon has long been known, and is called dispersion. One result of it is that it is not easy to say at what rate waves a few inches or a few yards or miles long ought to travel, by merely knowing at what rate the ultra-microscopic light-waves travel.

147. But there is even more to be said than this. There is not only dispersion, there is selective absorption possessed by matter : not only does it transmit different-sized waves at different rates, but it absorbs and quenches some much faster than others. Few substances, perhaps none, are equally transparent to all sizes of waves. Glass, for instance, which transmits readily the assortment of waves able to affect the retina, is practically quite opaque to waves two or three times longer or shorter. And whenever this selective absorption occurs, the laws of dispersion are extraordinary—so extraordinary that the dispersion is often spoken of as " anomalous "; which of course

means, not that it is lawless, but that its laws are unknown. Dispersion in any case is an obscure and little understood subject, but dispersion modified by selective absorption is still worse. Until the theory of dispersion is better understood, no one is able to say at what speed waves of any given length ought to travel. One can only examine experimentally at what rate they *do* travel. This has been done for long electrical waves, and it has been done for short light-waves : in the case of some substances the speed is the same, in the case of others it is different. But that the speed should be different is as I have now explained very natural, and can by no means be twisted into an admission that light-waves and electrical waves are not essentially identical. That the speed of both should agree at all is noteworthy ; the agreement appears to be exact in air, and practically exact in such simple substances as sulphur, and in the class of hydrocarbons known as paraffins ; whereas in artificial substances like glass, and in organic substances like fats and oils, the agreement is less perfect.

148. So much for the vital question of the speed at which electrical and optical disturbances travel. In some cases the speeds are accurately the same, in no case are they entirely different ; and in those cases where the agreement is only rough, an efficient and satisfactory explanation of the difference is to hand in the very different lengths of wave which have at

present been submitted to experiment. To compare the speeds properly, we must either learn to shorten electrical waves, or to lengthen light-waves, or both, and then compare the two things together when of the same size. It cannot be seriously doubted that they will turn out identical.

Manufacture of Light.

149. The conclusions at which we have arrived, that light is an electrical disturbance, and that light-waves are excited by electric oscillations, must ultimately, and may shortly, have a practical import.

Our present systems of making light artificially are wasteful and ineffective. We want a certain range of oscillation, between 7000 and 4000 billion vibrations per second : no other is useful to us, because no other has any effect on our retina ; but we do not know how to produce vibrations of this rate. We can produce a definite vibration of one or two hundred or thousand per second ; in other words, we can excite a pure tone of definite pitch ; and we can command any desired range of such tones continuously by means of bellows and a keyboard. We can also (though the fact is less well known) excite momentarily definite ethereal vibrations of some million per second, as I have explained at length; but we do not at present seem to know how to maintain this rate quite continuously. To get much faster rates of vibration than this we have to fall back upon atoms. We know how to make atoms vibrate : it is done by what we call " heating " the substance ; and if we could deal with individual atoms unhampered by others, it is possible that we might get a pure and simple mode of vibration from them. It is possible, but unlikely ; for atoms, even when isolated, have a multitude of modes of vibration special to themselves, of which only a few are of practical use to us, and we do not know how to excite some without also the others. However we do not at present even deal with individual atoms ; we treat them crowded together in a compact mass, so that their modes of vibration are really infinite.

We take a lump of matter, say a carbon filament or a piece of quick-lime, and by raising its temperature we impress upon its atoms higher and higher modes of vibration, not transmuting the lower into the higher but superposing the higher upon the lower, until at length we get such rates of vibration as our retina is constructed for, and we are satisfied. But how wasteful and indirect and empirical is the process. We want a small range of rapid vibrations, and we know no better than to make the whole series leading up to them. It is as though, in order to sound some little shrill octave of pipes in an organ, we were obliged to depress every key and every pedal, and to blow a young hurricane.

S

150. I have purposely selected as examples the more perfect methods of obtaining artificial light, wherein the waste radiation is only useless, and not noxious. But the old-fashioned plan was cruder even than this, it consisted simply in setting something burning : whereby not only the fuel but the air was consumed, whereby also a most powerful radiation was produced, in the waste waves of which we were content to sit stewing, for the sake of the minute, almost infinitesimal, fraction of it which enabled us to see.

Everyone knows now, however, that combustion is not a pleasant or healthy mode of obtaining light ; but everybody does not realize that neither is incandescence a satisfactory and unwasteful method which is likely to be practised for more than a few decades, or perhaps a century.

Look at the furnaces and boilers of a great steam-engine driving a group of dynamos, and estimate the energy expended ; and then look at the incandescent filaments of the lamps excited by them, and estimate how much of their radiated energy is of real service to the eye.   It will be as the energy of a pitch-pipe to an entire orchestra.

It is not too much to say that a boy turning a handle could, if his energy were properly directed, produce quite as much real light as is produced by all this mass of mechanism and consumption of material.

151. There might, perhaps, be something contrary to

the laws of Nature in thus hoping to get and utilize some specific kind of radiation without the rest, but Lord Rayleigh has shown in a short communication to the British Association at York [1] that it is not so, and that therefore we have a right to try to do it.

We do not yet know how, it is true, but it is one of the things we have got to learn.

Anyone looking at a common glow-worm must be struck with the fact that not by ordinary combustion, nor yet on the steam-engine and dynamo principle, is that easy light produced. Very little waste radiation is there from phosphorescent things in general. Light of the kind able to affect the retina is directly emitted, and for this, for even a large supply of this, a modicum of energy suffices.

Solar radiation consists of waves of all sizes, it is true; but then solar radiation has innumerable things to do besides making things visible. The whole of its energy is useful. In artificial lighting nothing but light is desired; when heat is wanted it is best obtained separately, by combustion. And so soon as we clearly recognize that light is an electrical vibration, so soon shall we begin to beat about for some mode of exciting and maintaining an electrical vibration of any required degree of rapidity. When this has been accomplished, the problem of artificial lighting will have been solved.

[1] *B. A. Report*, 1881, p. 526.

Mechanism of Electrical Radiation.

152. In forming a mental image of an electrical
wave, we have to note that three distinct directions are involved. There is (1) the direction of
propagation—the line of advance of the waves;
(2) the direction of the electric displacements, at right
angles to this; and (3) the direction of the magnetic
axis, at right angles to each of the other two.

One may get a rough mechanical idea of the
process of electrical radiation (at any rate in a plane)
by means of the cog-wheel system already used in
Part III. Imagine a series of elastic wheels, in one
plane, all geared together, and let one of them be
made to twist to and fro on its axis; from it, as
centre, the disturbance will spread out in all directions, each wheel being made to oscillate similarly
and to transmit its oscillation to the next. Looking
at what is happening at a distance from the source,
we shall see the pulses travelling from left to right;
the electrical displacement, such as it is, being up
and down; and the oscillating axes of the wheels
being to and fro, or at right angles to the plane
containing the wheels. The electric displacement
is small, because the positive and negative wheels
gearing into one another move almost equally, and
accordingly there is the merest temporary balance

of one above the other, due to the elastic " give " of the wheels. The magnetic oscillations, on the other hand, are all in one sense, the positive wheels rotating one way and the negative the other : all act together, and so the magnetic oscillation is a more conspicuous fact than the electric oscillation. Hence it is often spoken of as electro-magnetic radiation rather than as electric radiation. But the energy of the electrostatic strain is just as great as that of the electro-magnetic motion ; in fact the energy alternates from the potential to the kinetic form, or *vice versâ*, at every quarter swing, just like every other case of vibration.

153. As a matter of fact the magnetic oscillations are very small too. For just consider that the wheel-work extends right away to infinity in all directions : how is any moderate force going to make one of these wheels oscillate ? If they were rigid it would be impossible, but as they are elastic it is possible, though only with a very small amplitude of vibration ; and it sets up a strain all round which rapidly spreads out as we have said in all directions from the source. If the source were inclosed in a perfect conductor of moderate dimensions—if, for instance, one tried to oscillate one of the wheels inside the empty contour of Fig. 38—it would be easy enough : the wheels are limited in number, and can be easily got to oscillate considerably by a feeble source of disturbance.

This is commonly spoken of as the concentration of
light by reflection ; the conductor is said to act as a
perfect mirror ; and, since none of the light escapes,
any amount of illumination can be produced inside a
closed spherical mirror of perfect conductivity. Such
illumination would not be much use, however ; for,
directly a bit of matter is introduced to receive the
benefit of it, dissipation goes on at its surface, and
the violence of the ethereal disturbance is brought
down to something more moderate. Nevertheless,
even when dissipation is allowed, and when the
reflecting surface is by no means perfectly conducting, but is bright silver, which is the best conductor
we know, a considerable increase in illumination is
caused—by reflection, if we choose to say so—by
limitation, in at least some directions, of the extent of
ethereal medium to be affected by a given source, as
we might now prefer to express it (§ 164).

154. Prof. Fitzgerald, of Dublin, has devised a
model of the ether, which by help of a little artificiality represents the two kinds of displacement—
the electric and magnetic—very simply and clearly.

His wheels are separated from one another by a
certain space, and are geared together by elastic
bands. They thus turn all in one direction, and
no mention need be made of positive and negative
electricity as separate entities.

But, the wheels being massive, a rotatory disturbance given to one takes time to spread through the series, at a pace depending on the elasticity of the bands and the inertia of the wheels ; and during the period of acceleration one side of every elastic is stretched, while the other side is relaxed and therefore thickened. This thickening of the elastics goes on in one direction, and corresponds to an electric displacement in that direction ; the direction being

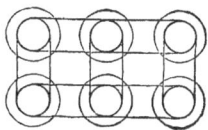

FIG. 48.—Fitzgerald's Ether Model. A set of massive-rimmed brass wheels on fixed pivots with their axles connected by common elastic bands. If the bands are taken off any region, it becomes a perfect conductor, into which disturbances cannot penetrate. (Cf. Fig. 38.)

perpendicular both to the direction of advance of the disturbance and to the axes of the wheels. A row of wheels corresponds to a section of a wave-front ; the displacements of india-rubber and the rotating axes, *i.e.* the electric and the magnetic disturbances, both lie in the wave-front.

155. Clerk Maxwell's originally suggested representation was not unlike this.[1] It consisted of a series of massive wheels, connected together not by a series of elastic bands but by a row of elastic particles or " idle wheels." These particles represented " electricity" ; their displacement during the period of ac-

[1] *Phil. Mag.*, April 1861.

celeration corresponding to the one-sided thickening of the elastic bands in Fitzgerald's model.

I have proposed to contemplate a double series of wheels geared directly into one another, and representing positive and negative electricity respectively, because it seems to me that so many facts point to the existence of these two entities, and because then no distinction has to be drawn between one part of the medium which is ether, and another part which is electricity, but the whole is ether and the whole is also electricity, while, nevertheless, a much-needed distinction can be drawn between a motion of the ether as a whole, and a relative motion of its component parts— a distinction between forces able to move ether, *i.e.* to displace the centre of gravity of some finite portion of it, and forces which shear it and make its components slide past each other in opposite senses : these latter forces being truly electromotive (§ 120).

156. If it be asked how the elasticity of the ether is to be explained, we must turn to the vortex sponge theory, suggested by Mr. Hicks[1] (Principal of Firth College, Sheffield), and recently elaborated by Sir William Thomson.[2] But this is too complicated a matter to be suited for popular exposition just at present. It must suffice to indicate that the points here left unexplained are not necessarily at the present

---

[1] *Brit. Assoc. Report*, 1885, Aberdeen, p. 930.
[2] *Ibid.* 1887, Manchester, p. 486. Also *Phil. Mag.*, October 1887.

time unexplainable, but that the explanations have not yet been so completely worked out that an easy grasp can be obtained of them by simple mechanical illustrations and conceptions. At the same time, the general way in which motion is able to simulate the effects of elasticity will be found popularly illustrated by help of gyrostats in Sir William Thomson's article "Elasticity" in the *Encyclopædia Britannica;*[1] and the fact that elastic rigidity of a solid can be produced by impressing motion on a homogeneous and otherwise structureless fluid must be regarded as one of the most striking among his many vital discoveries.

We have found it necessary all through Part III. to imagine the ether as composed of cells containing electricity in rotation, and that the act of magnetization consisted in facing these whirls round. Sir William Thomson has taught us that a medium containing whirls like this will simulate the behaviour of an elastic solid, and in fact that whirling motion is all that is required to explain elasticity (Fig. 46). With this hint, which might be developed at greater length, I must leave this part of the subject.

157. We have seen that to generate radiation an electrical oscillation is necessary and sufficient, and we have attended mainly to one kind of electric oscillation, viz. that which occurs in a condenser circuit when

---

[1] Also in a recent volume of the Nature Series—*Popular Lectures and Addresses.*

the distribution of its electricity is suddenly altered—as, for instance, by a discharge (§§ 124, 141). But the condenser circuit need not be thrown into an obviously Leyden jar form ; one may have a charged cylinder with a static charge accumulated mainly at one end, and then suddenly released. The recoil of the charge is a true current, though a weak one ; a certain amount of inertia is associated with it, and accordingly oscillations will go on, the charge surging from end to end of the cylinder like the water in a tilted bath suddenly levelled.

In a spherical or any other conductor, the like electric oscillations may go on : and the theory of these oscillations has been treated with great mathematical power both by Mr. Niven and by Prof. Lamb.[1]

Essentially, however, the phenomenon is not distinct from a Leyden jar or condenser circuit, for the ends of the cylinder have a certain capacity, and the cylinder has a certain self-induction ; the difficulty of the problem may be said to consist in finding the values of these things for the given case. The period of an oscillation may still be written $2\pi\sqrt{(LS)}$; only, since L and S are both very small, the "frequency" of vibration is likely to be excessive. And when we come to the oscillation of an atomic charge the frequency may easily surpass the rate of vibration

---

[1] *Phil. Trans.*, 1881 and 1883. Also by Prof. J. J. Thomson, *Math. Soc. Proc.*, April 1884.

which can affect the eye. The damping out of such vibrations, if left to themselves, will be also a very rapid process, because the initial energy is but small. It can be calculated that the oscillation of an atomic charge would give rise to only ultra-violet rays. It is probably because these ultra-violet rays synchronize with the period of vibration of atomic charges that they have such extraordinarily powerful chemical effects (§ 187).

158. But whether the charge oscillates in a stationary conductor, or whether a charged body vibrates as a whole, it equally constitutes an alternating current, and can equally well be treated as a source of radiation. Now, when we were considering the subject of electrolysis, we were led to think of molecules as composed of two atoms or groups of atoms, each charged with equal quantities of opposite kinds of electricity. Under the influence of heat the components of the molecules are set in vibration like the prongs of a tuning-fork, the rate of vibration depending on and being characteristic of the constants of the particular molecule. The atoms being charged, however, their mechanical oscillation is necessarily accompanied by an electric oscillation, and so an electric radiation is excited and propagated outwards. These vibrations would appear to be often of the frequency suited to our retina, hence these vibrating atoms indirectly constitute our usual source of light. The " frequency "

of the visible radiation can be examined and determined by optical means (some form of interference
experiment, usually a diffraction grating), and hence
many of the rates of vibration possible to the atoms
of a given molecule under given circumstances become
known, and this is the foundation of the science of
spectroscopy.

It is possible that the long duration of some kinds
of phosphorescence may be due to the atoms receiving
indirectly some of the ethereal disturbance, and so
prolonging it by their inertia, instead of leaving it to
the far less inertia of the ether alone. It is possible
also that the definite emissivity of some fluorescent
substances is due to periods of vibration proper to
their atoms, which, being disturbed in an indirect way
by receipt of radiation, re-emit the same radiation in
a modified, and, as it were, laden manner.

159. To get some further idea concerning the way in
which an oscillating charge or an oscillating charged
body can propagate radiation, refer back to Fig. 39,
Part III., and imagine the rack oscillating to and fro.
It will produce rotatory oscillation in the wheels
gearing into it, these again in the next, and so on. If
the wheel-work were rigid, the propagation would go
on at an infinite speed to the most distant wheels, but
if it be elastic then the pace of propagation depends
on the elasticity and the density in a way we have
already said enough about. The line of rack is the

direction of electric oscillation, the axes of the wheels the direction of magnetic rotatory oscillation, and at right angles to both these is the direction of advance of the waves. True, the diagram is not a space representation, it is a mere section, and a very crude suggestion of a mechanical analogy to what may be taking place.

The wheels being perfectly geared together and into the rack represent an insulator or dielectric: there is no slip or frictional dissipation of energy—in other words, there are no true electric currents. The electric oscillation is a mere displacement oscillation due to elasticity and temporary give of the elastic wheels, whereby during each era of acceleration they are thrown slightly into the state represented in Fig. 46 as contrasted with Fig. 37.

Effects of encountering a New Medium.

160. Now contemplate an advancing system of waves, and picture their encounter with an obstacle ; say, a medium of greater density, or less elasticity, or both. If the new medium is a perfect insulator, it must be considered as having its wheels thoroughly geared up both with themselves and with those of the initial medium, so that there is no slip or dissipation of energy at the surface. In this case none of the radiation will be lost : some will be reflected

and some transmitted according to ordinary and well-known mechanical laws. The part transmitted will suddenly begin to travel at a slower pace, and hence if the incidence were oblique would pursue a somewhat different path. Also, at the edges of the obstacle, or at the boundary of any artificially limited portion of the wave, there will be certain effects due to spreading out and encroaching on parts of the medium not lying in the direct path. These refraction and diffraction effects are common to all possible kinds of wave propagation, and there is nothing specially necessary to be said concerning electrical radiation on these heads which is not to be found in any work on the corresponding parts of optics.

161. Concerning the amount and direction of the reflected vibrations there is something to be said however, and that something very important ; but it is no easy subject to tackle, and I fear must be left, so far as I am concerned, as a distinct, but perhaps subsequently-to-be-filled-up, gap.

If the gearing between the new medium and the old is imperfect, if, for instance, there were a layer of slippery wheels between them, representing a more or less conducting film, then some of the radiation would be dissipated at the surface, not all would be reflected and transmitted, and the film would get to a certain extent heated. By such a film the precise laws of reflection might be profoundly modified, as they would be also if the transition from one medium to another were gradual instead of abrupt. But all these things must remain for the present part of the unfilled gap.

Electric Radiation encountering a Conductor.

162. We will proceed now to the case of a *conducting* obstacle—that is, of waves encountering a medium whose electrical parts are connected, not by elasticity, but by friction. It is plain here that not only at the outer layer of such a medium, but at every subsequent layer, a certain amount of slip will occur during every era of acceleration, and hence that in penetrating a sufficient thickness of a medium endowed with any metallic conductivity the whole of the incident radiation must be either reflected or destroyed : none can be transmitted (§ 104).

Refer back to Fig. 43, and think of the rack in that figure as oscillating. Through the cog-wheels the disturbance spreads without loss, but at the outer layer of the conducting region A B C D a finite slip occurs, and a less amount of radiation penetrates to the next layer, E F G H, and so on. Some thickness or other, therefore, of a conducting substance must necessarily be impervious to electric radiation : that is, it must be opaque. And since it dissipates very little energy, it must act as a reflector. (See §§ 153 and 164.)

Conductivity is not the sole cause of opacity.   It
would not do to say that all opaque bodies must be
conductors.   But conductivity is a very efficient cause
of opacity, and it is true to say that all conductors of
electricity are necessarily opaque to light; understanding, of course, that the particular thickness of
any homogeneous substance which can be considered
as perfectly opaque must depend on its conductivity.
It is a question of dissipation, and a minute but
specifiable fraction of an original disturbance may be
said to get through any obstacle.   Practically, however,
it is well known that a thin, though not the thinnest,
film of metal is quite impervious to light.

163. When one says that conductivity is not the sole
cause of opacity, one is thinking of opacity caused by
heterogeneity.   A confused mass of perfectly transparent substance may be quite opaque ; witness foam,
powdered glass, chalk, &c.

Hence, though a transparent body must indeed be
an insulator, the converse is not necessarily true.   An
insulator need not necessarily be transparent.   A
homogeneous flawless insulator must, however, be
transparent to some, though not necessarily to all
wave-lengths.   A homogeneous and flawless opaque
body, if really opaque to all wave-lengths, must be a
conductor.

These, then, are the simple connections between
two such apparently distinct things as conducting

power for electricity and opacity to light which
Maxwell's theory points out; and it is possible
to calculate the theoretical opacity of any given
simply-constructed substance by knowing its specific
electric conductivity.

Fate of the Radiation

164. To understand what happens to radiation impinging on a conducting body it is most simple to
proceed to the limiting case at once and consider a
perfect conductor. In the case of a perfect conductor
the wheels are connected not even by friction; they
are not connected at all. Consequently the slip at
the boundary of such a conductor is perfect, and there
is no dissipation of energy accompanying it. The
blank space in Fig. 38 represented a perfectly conducting layer. Ethereal vibrations impinging on a
perfect conductor practically arrive at an outer confine of their medium : beyond, there is nothing capable
of transmitting them ; the outer wheels receive an impetus which they cannot get rid of in front, and which
they therefore return back the way it came to those
behind them with a reversal of phase : the radiation is
totally reflected. It is like what happens when a
sound-pulse reaches the open end of an organ-pipe ;
like what happens when sound tries to go from water
to air; like the last of a row of connected balls along

T

which a knock has been transmitted ; and our massive elastic wheels, especially the wheels of Fig. 48, are able to represent the reversal of phase and reflection quite properly.

165. The reflected pulses will be superposed upon and interfere with the direct pulses, and accordingly if the distances are properly adjusted we can have the familiar formation of fixed nodes and stationary waves (§ 130).

166. The point of main interest, however, is to notice that a perfect conductor of electricity, if there were such a thing, would be utterly impervious to light: no light could penetrate its outer skin, it would all be reflected back : the substance would be a perfect reflector for ethereal waves of every size.

Thus with a perfect conductor, as with a perfect non-conductor, there is no dissipation. Radiation impinging on them is either all reflected or some reflected and some transmitted. It is the cases of intermediate conductivity which destroy some of the radiation and convert its ethereal vibrations into atomic vibrations, *i.e.* which convert it into heat.

167. The mode in which radiation or any other electrical disturbance diffuses with continual loss through an imperfect conductor can easily be appreciated by referring to § 103 again. The successive lines of slip, A B C D, E F G H, &c., are successive layers of induced currents. An electromotive impulse loses

itself in the production of these currents, which are successively formed deeper and deeper in the material according to laws of diffusion.

If the waves had impinged on one face of a slab, a certain fraction of them would emerge from the other face—a fraction depending on the thickness of the slab according to a logarithmic or geometrical-progression law of decrease.

CHAPTER XV.

168. WE must now mention one or two phenomena which depend entirely upon a modification of ether by the neighbourhood of matter, and which we have reason to believe would not occur in free ether at all. These are the optical phenomena of Faraday and Kerr, and the electric phenomenon of Hall.

Faraday discovered, long before there was any other connection known between electricity and light, that the plane in which light-vibrations occur could be rotated by transmitting light through certain magnet- ized substances along the lines of magnetic force. To make this effect easily manifest, one uses plane-polar- ized light, and transmits it through a fair length of magnetized substance, analyzing it after emergence and showing that, though it remains plane-polarized, the plane has been rotated, possibly through a right angle or more.

Now, in a general way it is easy to imagine that, inasmuch as something of the nature of a rotation is going on in a magnetic field round the lines of force, vibrations travelling into such a field along these lines should be twisted round, corkscrew fashion, and emerge vibrating in a different plane. But when one tries to follow out this process into detail, one finds it not quite so simple a matter. It has, however, no business to be a very simple and obvious consequence of the existence of a magnetic rotation round the rays of light, else would it occur in free space, and in the same direction in all media. But the facts are that in free space—that is, in free ether—it does not occur at all, and the direction of rotation is not the same for all media : substances can, in fact, be divided into two groups, according to the way in which given magnetization shall rotate the plane of polarized light passing through them.

169. Similar statements can be made concerning the electrostatic optical effect discovered by Dr. Kerr, who showed that plane-polarized light transmitted across the lines of force in an electrostatic field could, in certain media, come out elliptically polarized. Now, inasmuch as an electric field is a region of strain, and strain in transparent bodies is well known to make them slightly doubly refracting and able to turn plane-polarized into elliptically-polarized light, it is very easy to imagine such a result in an electric field to be

natural and probable. But the explanation is not so simple as that, else it ought to be a large effect, occurring in all sorts of media in the same direction, and likewise in free space. But the facts are that it does not occur at all in free space, and it occurs in different senses in different substances ; so that again they can be grouped into two classes according to the sign of the Kerr effect.

Thus, then, the rotatory effect of a magnetic field upon light, discovered by Faraday, and the doubly refracting effect of an electrostatic field upon light, discovered by Kerr, agree in this : that they are both small or residual effects, depending on the existence of a dense medium, and both varying in sign according to the nature of the medium.

170. The only substance in which the Faraday effect is large, is iron, including with iron the other highly magnetic substances. The discovery of the effect in these bodies was likewise made by Kerr. The difficulty of dealing with them is that they are very opaque, and hence that the merest film of them can be used. The film can be used either by way of transmission or by way of reflection, it matters not which, but reflection is the way in which it was first done. Light reflected from the polished face of a magnet has indeed barely penetrated at all into the substance of the iron before being sent back ; still, it has penetrated deep enough to be

distinctly rotated by the tremendous magnetic whirl which it finds there.

171. All these highly magnetic substances are metallic conductors, and are therefore very opaque. Whether there is any real connection between high magnetic susceptibility and conductivity is more than I can say. But it is quite natural, and indeed necessary, that the greatest portion of light should be reflected on entering a highly magnetic medium, because in such a medium the ethereal density, μ, is so great, and hence the velocity of wave transmission must undergo a sudden and immense decrease—a circumstance always causing a great amount of reflection, just as when sound tries to pass from any one medium to a much denser one.

But the opacity of iron and other magnetic substances may be explained by the mere fact of their conducting power, just like other metals, and no noteworthy effect of their large value of μ need be detectable optically.

If a non-conducting highly magnetic substance could be found, it would probably reflect a great deal of light at its surface, though it would not dissipate that which entered it. Such a substance would be most interesting to submit to experiment, but perhaps its existence presupposes a combination of impossible properties. Certainly it has not yet been discovered.

As to the phenomenon detected by Hall, it appears

intimately associated with that of Faraday, and it will be most simple to omit all reference to it for the present.

172. A general idea of what is happening in the Faraday and Kerr phenomena can be given thus.

A simple vibration, like a pendulum-swing, or any other oscillation in one plane, can be resolved into two others in an infinite variety of ways; just as one force can be resolved into any number of pairs of equivalent forces. The two most useful modes of

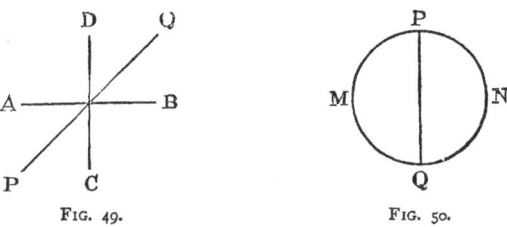

FIG. 49.   FIG. 50.

analyzing a simple vibration into a pair of constituents are these : (1) two equal components, likewise plane vibrations, each inclined at 45° to the original one, as when P Q is resolved into A B and C D (Fig. 49) ; and (2) two equal circular or rotatory oscillations in opposite directions, as when P Q is resolved into P M Q and P N Q (Fig. 50). The first method of resolution is useful in explaining Kerr's effect, the second in explaining Faraday's.

Of the two component vibrations, A B and C D, into which P Q can be supposed analyzed, let some cause, no matter what, make one gain upon the other, so that in travelling along a line perpendicular to the paper one goes a little the quicker : the effect at once is to change the character of the vibration into which they will recompound.   After the gain, they no longer reproduce the original simple vibration P Q : they give rise to elliptic, or it may be to circular, vibrations ; this last, if the retardation is equal to a quarter period.

These are matters fully treated in any elementary treatise on polarized light, and they are quite easily illustrated by means of a simple pendulum.   One may assume them known.

Similarly with the second system of analyzing the vibration into two opposing circular ones.   If the components travel through any interposed medium at the same rate, they will, on emergence, reproduce the original vibration in its original position ; but if one travels quicker than the other they recombine into a vibration of the same character as at first, but turned through a certain angle.   Thus anything which retards one of the *rectangular* components behind the other, changes the character of the vibration from plane into elliptical ; while anything which retards one of the *circular* components behind the other, leaves the character of the vibration unaltered, but rotates it through a certain angle.

173. So far one has said nothing but the simplest
mechanics. The next point to consider is what determines the rate at which light travels through any substance? This we have discussed at length (§ 128), and
shown to be $\dfrac{1}{\sqrt{(K\mu)}}$. Anything which increases either
the electric or the magnetic permeability of the
medium decreases the velocity of light. Now, when
a medium is already subject to a violent strain in any
one direction it is possibly less susceptible to further
strain in that direction and responds less readily. Not
necessarily so at all : such an effect would only be
produced when the strain was excessive, when the
medium was beginning to be overdone, and when its
properties began thereby to be slightly modified.
There are reasons for believing the specific inductive
capacity of most media to be very constant ; of some
media, perhaps, precisely constant ; but if there were
any limit beyond which the strain could not pass it is
probable that on nearing the limit the specific inductive capacity would be altered—possibly increased,
possibly diminished—one could hardly say which.
Quincke has investigated this matter, and has shown
that the value of K is affected by great electric strain.

Suppose now that a dielectric is subject to a violent
electrical stress, so that its properties along the lines
of force become slightly different from its properties
at right angles to those lines. The value of K will

not be quite the same along the lines of strain as across
them, and accordingly the rectangular component of
a vibration resolved along the lines of force will travel
rather quicker or rather slower than the component at
right angles, because the velocity of transmission
depends upon K, as already explained: such a medium at once acquires the necessary doubly-refractive
character, and will show Kerr's effect.

174. Similarly with magnetization.   It is well
known that for many media μ is not constant.   Take
iron, for instance.   For very small magnetizing forces
the susceptibility is moderate, and increases as they
increase ; at a certain magnetization it reaches a maximum, and then steadily decreases.   But not only is
it thus very inconstant, its ascending and descending
values are not the same.   To forces tending to magnetize it more, the susceptibility has one value ; to
forces tending to demagnetize it, it has another and
in general smaller value.   This property has been
specially studied by Ewing, and has been called by
him "hysterēsis."   Slightly susceptible substances
cannot be magnetized to anything like the same
extent, and hence the property in them has been less
noticed, perhaps not noticed at all.   Nevertheless it
must exist in every substance which exhibits a trace
of permanent magnetism, and every substance I have
tried appears to show some such trace.[1]

[1] See *Nature*, vol. xxxiii. p. 484.

An already strongly magnetized medium will be rather differently susceptible to additional magnetizing forces in the same direction from what it is to those in a contrary direction. Nothing more is wanted to explain Faraday's effect. The vibration being resolved into two opposite circular components, one of them must agree in direction with the magnetism already in the medium and try to magnetize it for the instant infinitesimally more; the other component will for the instant infinitesimally tend to demagnetize it. The value of μ offering itself to the two components will be different, hence they will go at different rates, and the plane of vibration will be rotated.[1]

175. The direction of rotation will depend on whether the value of μ is greater for small relaxations, or for small intensifications, of magnetizing force; and diamagnetic substances may be expected

---

[1] The connection which I here trace between hysterēsis and the magnetic rotation of plane of polarization, is not one which I at all press. Prof. Fitzgerald has intimated to me that if I take a whole wave-front into consideration, the theory will hardly work, and that it would have been better if the real electro-magnetic disturbance were the thing acted on instead of having to fall back upon a secondary magnetic effect of the electrostatic displacement. And Prof. Ewing, though he adduced at first some facts which appeared to strengthen my view, now doubts whether the kinematic resolution of a displacement into two circular components is under the circumstances legitimate. I have doubts too. If I were quite sure that there were no vestige of truth in the suggestion I have made in the text, I should of course suppress it ; but as I am not quite sure, I let it stand for the present, taking any possible harm out of it by this note.   See also §§ 180 *et seq.*

to be opposite in this respect to paramagnetic ones. Any substance for which μ is absolutely constant, whatever the strength of magnetic polarization to which it is submitted, can hardly be expected to exhibit any hysterēsis; the ascending and descending curves of magnetization will coincide, being both straight lines, and such a substance will show no Faraday effect. Similarly, any substance for which K is absolutely constant, whatever the electric polarization to which it is submitted, can show no Kerr's effect. Free space appears to be of this nature; and gases approach it very nearly, but not quite.

In iron, μ is greater for an increasing than for a decreasing force, as is shown by the loops in Ewing's curves; hence the circular component agreeing in direction with the magnetizing current will travel slower than the other component, and hence the rotation in iron will be against the direction of the magnetizing current. The same appears to hold in most paramagnetic substances, and the opposite in most diamagnetic; but the mere fact of paramagnetism or diamagnetism is not sufficient to tell us the sign of the effect in any given substance. We must know the mode in which its magnetic permeability is affected by waxing and by waning magnetization respectively.

Possible Electrical Method of detecting the Faraday Effect.

176. Thus far we have considered the rotation of electric displacement by a magnetic field as being examined optically, the displacements being those concerned in light, and the rotation being detected by a polarizing analyzer suitable for determining the direction in which the vibrations occur before and after the passage of light through a magnetized substance. This is the only way in which the effect has at present been observed in transparent bodies. But one ought not to be limited to an optical method of detection.

Electrical displacements are easily produced in any insulator, and if it be immersed in a strong magnetic field, so that the electric and magnetic lines of force are at right angles to each other, every electric disturbance ought to experience a small rotation. A steady strain will not be affected ; it is the variable state only which will experience an effect, but every fresh electric displacement should experience a slight rotatory tendency just like the displacements which occur in light.

Now to rotate a displacement A B into the position A C requires the combination with it of a perpendicular displacement B C (Fig. 51). Hence the effect of the magnetic field upon an electric displacement, A B, may be said to be the generation of a small perpendicular E.M.F., B C, which, compounded with the original one, has the resultant effect A C.   It will be

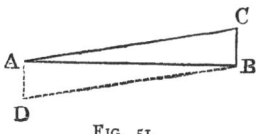

FIG. 51.

only a temporary effect, lasting while the displacement is being produced, and ceasing directly a steady state of strain is set up.

An inverse E.M.F., A D, will be excited by the same magnetic field directly the displacement is reversed.

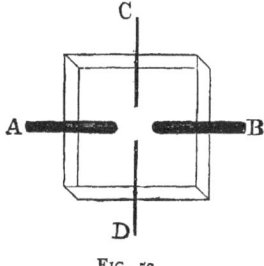

FIG. 52.

And so, if a continual electric oscillation is kept up between A and B in a magnetic field, an accompanying very minute transverse oscillation may be expected, and may be looked for electrically.

Some such arrangement as that here shown (Fig. 52) may be employed : a square of heavy glass, perforated with four holes towards the centre, supplied with electrodes : one pair of electrodes, A, B, to be connected with the poles of some alternating machine, and the other pair, C, D, connected to a telephone or other detector of minute oscillatory disturbance. So soon as a strong steady magnetic field is applied, by placing the glass slab between the poles of a strong magnet, the telephone ought to be slightly affected by the transverse oscillations. This effect has not yet been experimentally observed, but it seems to me a certain consequence of the Faraday rotation of the plane of polarization of light.

Hall Effect.

177. Although the existence of this transverse E.M.F., excited by a magnetic field in substances undergoing varying electric displacement, has at present only been detected optically in transparent bodies, *i.e.* in insulators, yet in conductors the corresponding effect with a steady current has been distinctly observed electrically. By many persons it had been looked for (by Prof. Carey Foster and the writer, among others, though unfortunately they were not sufficiently prepared for its extreme smallness); by Mr. Hall, at Baltimore, was it first successfully observed.

In conductors it is natural to use a conduction-current instead of a displacement-current. A steady current can be maintained in a square or cross of gold-leaf or other thin sheet of metal between the electrodes A, B ; and a minute transverse E.M.F. can be detected, causing a very weak steady current through a galvanometer connected to the terminals C, D, so soon as a strong magnetic field is applied perpendicularly

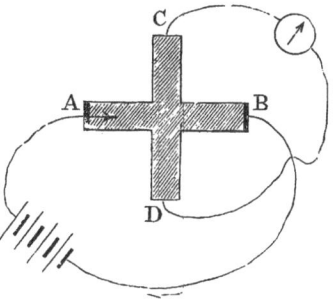

Fig. 53.—The direction of the transverse E.M.F. excited by the earth's vertical magnetic field in this conductor, conveying a current as shown, is C D if it represents gold, D C if it represents iron.

to the plate. Fig. 53 will sufficiently indicate the arrangement. The poles of the magnet are one above and one below the paper.

In iron it is easy to see which way the transverse E.M.F. ought to be found. It has been shown (§ 175) that a displacement will be rotated in iron against the magnetizing current ; hence to rotate the displacement A B to A C (Fig. 51), requires in iron a clockwise

U

magnetizing current. Such a current, or, what is the same thing, a south pole below the paper, a north pole above, excites, in the cross of Fig. 53, an E.M.F. in the direction D C ; and this by Ampère's rule is just the direction in which the conductor itself is urged by the magnetic forces acting on the current-conveying substance. Most diamagnetic substances should exhibit a transverse E.M.F. in the opposite sense. This transverse E. M. F. excited in conductors conveying a current in a magnetic field is the effect known by the name of Hall. It is, as Prof. Rowland and others have pointed out, intimately connected with the Faraday rotation of light.

178. Unfortunately a pure and simple Hall effect is a difficult thing to observe. Magnetism affects the conductivity of metals in a rather complicated manner ; and strain affects their thermo-electric properties. Now a metal conveying a current in a magnetic field is certainly more or less strained by mechanical forces ; and hence heat will be developed unequally in different parts, by a sort of Peltier effect ; and the result of this will be to modify the resistance in patches, and so to produce a disturbance of the flow which may easily result partly in a transverse E.M.F. This has been pointed out by Mr. Shelford Bidwell.

The more direct effect of magnetism on conductivity may be negligibly small in many metals, but in bismuth it is certainly large. Both of these spurious effects seem to be large in bismuth, and probably quite mask any true Hall effect there may be in that metal. In all cases the existence of these spurious effects makes it difficult to be sure of the magnitude and sign of the real rotational effect.

179. But, it may be asked, what right have we to distinguish between a real and a spurious Hall effect? If a transverse E.M.F. can be predicted by reason of known strains and thermo-electric properties, as well as by known rotation of light effects, why should the two things be considered different? Why should they not be different modes of regarding one and the same phenomenon?

In other words, may not the Faraday rotation of light vibration be due to infinitesimal temporary strains and heatings in the medium, caused by the fact that minute electric displacements are occurring in a violent magnetic field? This is a question capable of being answered by a quantitative determination of the amounts and direction of the effects to be expected, and a comparison with those actually observed. I do not know of data, at present obtained, sufficient to enable us to answer it. If its answer should turn out in the affirmative, several apparently distinct phenomena will be linked together.

Possible Accounts of the Faraday and Hall Effects.

180. The account I have given of the magnetic rotation of the plane of polarization has made it depend on the phenomenon of hysterēsis, in a way which may be thus summarized. The value of μ for increasing magnetization is different from that for decreasing magnetization ; an electric displacement such as occurs in every half-swing of a light-vibration is resolvable into two opposite circular components, one of which increases, while the other decreases, any magnetization already existing in the direction of the ray : the value of μ affects the speed of transmission of light ; hence the two circular components will not proceed at the same pace, and the direction of vibration will infinitesimally rotate. The same thing is repeated at every half-swing, the elemental rotations being all in the same sense, and so the ultimate rotation of the plane of polarization in transparent bodies is accounted for.

The Hall effect observed in conductors follows at once ; for the rotation of a displacement is equivalent to combining it with a small perpendicular displacement ; and it is this perpendicular or transverse E.M.F. exerted by a magnetic field which Hall discovered. At the same time, there are one or two facts which militate against this view of the Hall effect, chief among which is the singular behaviour of nickel, which rotates light one way and electric displacement the other way. For some time it was possible to hope for a way out of this through the usual convenient avenue of " impurity " ; but now that both experiments have been performed on the same identical piece of metal, still with opposite results, this exit is closed.   In this unsettled state, so far as I know, the connection between the rotation of light and the Hall effect at present stands.

181. It may be well here to repeat the caution appended as a footnote to § 174, not to assume that this account of the magnetic rotation of light and the Hall effect is true.   If true, however, it is convenient as linking the phenomena on to hysteresis, and the direction of the effect in iron is correctly given—namely, a rotation against the magnetizing current (§ 177).

Prof. Ewing has since pointed out, in a letter to me, that, attending more precisely to the instruction of his curves, we find the difference in μ for positive and negative magnetizing forces only lasts through a number of cycles for the time during which the final state has been approached, and does not persist after a steady state has been reached. This would make the magnetic rotation of light a function of time ; and certain experiments by Villari on spinning a glass disk between the poles of a magnet, so that fresh and fresh portions of glass were continually exposed to the magnetic field, showed a marked falling off in the amount of rotation as soon as high speeds were obtained ; thus proving, apparently, that a certain short time was necessary to set up the effect. This experiment, and other modifications of it, want repeating, however.[1]

Prof. Ewing has subsequently expressed a doubt as to whether the kinematic resolution of a displacement into two equal opposite circular components is, under the circumstances, legitimate.

Prof. Fitzgerald has further pointed out that, although when attending to one element only the theory might possibly work, yet, as soon as one takes into account the whole wave-front, it breaks down ; for all the main magnetic disturbance lies in the wave-front, as is well known, and the extra magnetic disturbance which I have postulated as a consequence of electrostatic displacement is annulled by interference of adjacent elements.

If I were quite sure that there were no vestige of truth in the suggestion I have made, I should, of course, withdraw it ; but, as I do not feel perfectly sure either way, I leave it in a dilapidated condition for the present.

182. Another and apparently distinct account of the magnetic rotation can also be hinted at, which links the phenomenon on to the facts of thermo-

----

[1] See in this connexion a paper by the present writer in the *Philosophical Magazine* for April, 1889. Also a paper by Mr. A. W. Ward, to be read before the Royal Society shortly.

electricity.   It labours under worse disadvantages
than the preceding, being more hazy.

Referring back to § 63, we find that, to explain
what is called the "Thomson effect" in metals, we
were led to suppose a connection between one kind
of electricity and some kinds of matter more
intimate than between the other kind of electricity and the same matter.   Thus, the atoms
of iron were said to have a better grip of positive
electricity than of negative ; while copper, on
the other hand, had a better grip of negative
than of positive.   All metals could be arranged in
one or other of the two classes, with the exception
of lead, which appears to grip both equally.   It is
the same phenomenon as was originally named by
Sir W. Thomson "the specific heat of electricity in a
substance."   Certain it is that vibrating atoms of iron
push positive electricity from places of more rapid, to
places of less rapid, vibration—that is, from hot to cold
—and a whole class of the metals do the same ; while
another class, like copper, push it from cold to hot.

Permitting ourselves to picture this effect as a direct
consequence of the Ohm's law relation between electricity and matter (§ 60), combined with a special
relationship between certain kinds of matter and one or
other kind of electricity, a relationship which can exhibit
itself in other ways also, we get a possible though rather
hazy notion of a Faraday rotation in a magnetic field

by supposing that the Amperian molecular currents in these substances consist not of precisely equal positive and negative currents, but of opposite currents slightly unequal ; say, for instance, that the density of the positive constituent of the bound ether of a substance is slightly different from that of the negative constituent ; so that on the whole the bound ether in a magnetized molecule is slowly rotating one way or the other, at a pace equal to the resultant rotation of its constituents. Suppose that in iron the positive Amperian electric current is the weaker of the two ; then the ether, as a whole, will be rotating with the negative current, and accordingly an ethereal vibration entering such a medium will begin to screw itself round in a direction opposite to that of the magnetizing current ; whereas in copper or other such substance it would be rotated the other way.

183. According to this (admittedly indistinct) view, lead ought to show no rotatory effect at all ; and of course, therefore, no Hall effect either. And the classes into which metals are divided by the sign of their Hall effect should coincide with the classes into which the sign of their Thomson effect throws them.

Hall finds that, of the metals he examined, iron, cobalt, and zinc fall into one class, while gold, silver, tin, copper, brass, platinum, nickel, aluminium, and magnesium fall into the other. Now, referring to the thermo-electric results of Prof. Tait, we find iron,

cobalt, platinum, and magnesium with a negative sign
to their Thomson-effect-coefficient, or with lines in the
thermo-electric diagram sloping downwards ; while
gold, silver, tin, copper, aluminium, and zinc slope
upwards, or have a positive sign to their " specific heat
of electricity."

According to this, therefore, the discordant metals
are zinc, platinum, and magnesium. The proper
thing to say under these circumstances is that the
metals used in the very different experiments were
not pure. They certainly were not ; but I do not feel
able to conscientiously bolster up so inadequate a
theory by help of this convenient fact.

In the *Philosophical Magazine* for May 1885, Mr.
Hall gives some more measurements, showing that in
bismuth the effect is enormous, and in the same direction as in copper, whereas in antimony it is also great,
and in the same direction as in iron. All these things
seem to point to some thermo-electric connection —
whether it be of the sort I have vaguely tried to
indicate, or some other.

Other Outstanding Problems.

184. Outstanding problems bristle all over the subject, and if I pick out any for special mention it will
only be because I happen to have made some experiments in their direction myself, or otherwise have had
my thoughts directed to them, and because they have

not been so directly called attention to in the body of the book.

Referring back to § 66 at the end of Part II., "a current regarded as a moving charge," it is natural to ask, Is this motion to be absolute, or relative to the ether only, or must it be relative to the indicating magnetometer ? In other words, if a charged body and a magnetic needle are flying through space together, as, for instance, by reason of the orbital motion of the earth, will the needle experience any deflecting couple ?

It is one of many problems connected with the ether and its motion near gross matter—problems which the experiments of Fizeau (showing that a variable part of ether was bound with matter and transmitted with it, while another constant portion was free and blew through it) began to throw light upon ; aberration problems such as have been partially solved by the genius of Stokes ; problems connected with the motion of ether near great masses of matter, like those which Michelson is so skilfully attacking experimentally : it is among these that we must probably relegate the question whether absolute or relative motion of electric charges is concerned in the production of magnetic field, and what absolute motion through the ether precisely means. It is doubtless a question capable of being attacked experimentally, but the experiments will be very difficult. I believe that Prof. Ayrton has attempted them.

185. Referring back to Parts I., II., and III., §§ 7, 39, 41, 48, 88, 89, 97, 98, 109, 122, 134, we find a number of questions regarding momentum left unsettled. Has an electric current any true momentum mechanically discoverable ? Now, this question before it can be answered in the negative, will have to be attacked under a great number of subdivisions. One may classify them thus. Two main heads : (1) When steady, Does a magnet behave in the least like a gyrostat ? (2) When variable, Is there a slight mechanical kick on starting or stopping a current ? With four or more subsidiary heads under each, viz. (a) in metallic conductors ; (b) in electrolytes ; (c) in gases ; (d) in dielectrics.

Suppose the answer turns out negative in metals, it by no means follows that it should be negative in electrolytes too. In fact, as matter travels with the current in the case of electrolytic conduction (§ 36), it is hardly possible that there is not some momentum, though it may be too small to observe—either a kick of the vessel as a whole at starting and stopping, or a continuous impact on an electrode receiving a deposit. The present writer has looked for these things, but after gradually eliminating a number of spurious effects the result has been so far negative. In a light quill vessel fixed to the end of a torsion arm, the main disturbance was due to variations of temperature which gradually introduced a minute air-bubble, and by kicking this backwards and forwards simulated the effects sought. In the case of the suspended electrode, convection currents in the electrolyte, caused by extra concentration or the reverse, seem determined to mask any possible effect.

One obvious though very troublesome source of disturbance in all cases is the direct effect of terrestrial magnetism on the circuit. To get over this, the writer not only made his circuits as nearly as possible of zero area, but also inclosed them in the iron case of a Thomson marine galvanometer, lent for the purpose by Dr. Muirhead.

In gases, the experiment of Mr. Crookes, where an electrical stream inside a vacuum-tube propels a mill along rails—perhaps even the ancient experiment of the blast from a point—shows that momentum is by no means absent from an electric current through a gas (§ 64).

To see if there are any momentum effects accompanying variation of electric displacement in dielectrics, the writer has suspended a mica-disk condenser at the end of a torsion arm, and arranged it so that it could be charged and discharged *in situ*. Many spurious effects, but no really trustworthy ones, were observed.

In the writer's opinion the subject is by no means thoroughly explored, and he only mentions his old attempts as a possible guide to future experimenters.

186. Then, again, there is the influence of light on conductivity. Annealed selenium, and perhaps a few other things, improve in conductivity enormously when illuminated. The cause of this is unknown at present, and whether it is a general property of matter, possessed by metals and other bodies to a slight degree, is uncertain ; for the experiments of Börnstein, with an affirmative result for the case of metals, have been seriously criticized.

Even though metals show no effect, yet electrolytes might possibly do so, but the effect, if any, is small ; and it is particularly difficult in their case to distinguish any direct radiation effect from the similar effect of mere absorbed radiation or heat.

The writer has found that a glass test-tube kept immersed in boiling water conducts distinctly better when the blinds of a room are raised than when they are lowered, though nothing but diffuse daylight falls upon it. But as the effect could have been produced by a rise in temperature of about the tenth of a degree, and as the absorption of diffuse daylight is competent to produce a rise of temperature as great as this in the glass of a thermometer-bulb even though immersed in boiling water, he feels constrained to regard the result, though very clear and distinct, as after all a negative one, and has accordingly not published it.

187. The fact that ultra-violet waves have a period

of vibration synchronous with probable electric vibration in molecules (§ 157) seems to cause a multitude of consequences now being discovered. Hertz noticed that the light of one spark influenced another at a distance, so that a sparking interval was virtually shortened when illuminated. Wiedemann and Ebert have further investigated this, and obtained several interesting results, distinctly proving that it is ultraviolet light which is effective. Hallwachs has discovered that a clean metallic plate becomes electrified when light falls upon it. And there are a number of other similar facts, some long known, some recent, which all illustrate the molecular effects of light. It appears probable that they all depend on some synchronized disturbance set up in the air or other film in contact with the substance, a disturbance resulting in some kind of chemical action, and hence that these physical effects are of the same order as those other familiar but vaguely grasped facts summed up under the category of the chemical or actinic power of light. For that light affects silver salts, ebonite, hydrogen and chlorine, &c., is an old story. Some progress is now likely to be made in ascertaining the precise mode in which these changes occur (§ 33).

188. A few months ago I should have put in a prominent position among outstanding problems the production of electric radiation of moderate wave-

length, and the performance with this radiation of all the ordinary optical experiments—reflection, refraction, interference, diffraction, polarization, magnetic rotation, and the like (§ 1). But a great part of this has now been done, and so these things come to be mentioned under a different heading :—

Conclusion.

"Conclusion" is an absurd word to write at the present time, when the whole subject is astir with life, and when every month seems to bring out some fresh aspect, to develop more clearly some already glimpsed truth. The only proper conclusion to a book dealing with electricity at the present time is to herald the advent of the very latest discoveries, and to prepare the minds of readers for more.

189. Referring back to Chap. XIV., to §§ 1 and 8, and all Part IV., we spoke confidently of a radiation being excited by electric oscillations, a radiation which travelled at the same rate as light, which is reflected and refracted according to the same laws, and which, in fact, is identical with the radiation able to affect our retina, except in the one matter of wave-length. Such a radiation has now been definitely obtained and examined by Dr. Hertz, of Karlsruhe ; and in the last month of last year, Prof. von Helmholtz communicated to the Physical Society of Berlin an account of Dr. Hertz's latest researches.

The step in advance which has enabled Dr. Hertz to do easily that which others have long wished to do, has been the invention of a suitable receiver. Light when it falls on a conductor excites first electric currents and then heat. The secondary minute effect was what we had thought of looking for; but Dr. Hertz has boldly taken the bull by the horns, looked for the direct electric effect, and found it manifesting itself in the beautifully simple form of microscopic sparks across a gap between two conductors, or between the ends of a looped conductor.

He takes a brass cylinder, some inch or two in diameter, and a foot or so long, divided into two halves with a small sparking interval between, and connects the halves to the terminals of a small induction-coil ; every spark of the coil causes the charge in the cylinder to surge to and fro about five hundred million times a second, and to disturb the ether in a manner precisely equivalent to a diverging beam of plane-polarized light, with waves about thrice the length of the cylinder.

The radiation, so emitted, can be reflected by plane conducting surfaces, and it can be concentrated by metallic parabolic mirrors ; the mirror ordinarily used being a large parabolic cylinder of sheet zinc, with the electric oscillator situate along its focal line. By this means the effect of the wave could be felt at a fair distance, the receiver consisting of a synchronized pair of straight conductors with a microscopic spark-gap between them, across which the secondary induced sparks were watched for. By using a second mirror like the first to catch the parallel rays and reconverge them to a focus, the effect could be appreciated at a distance of 20 yards, If the receiving mirror were rotated through a right angle, it lost its converging power on this particular light.

Apertures in a series of interposed screens proved that the radiations travelled in straight lines (roughly speaking, of course).

A gridiron of metallic wires is transparent to the waves when arranged with the length perpendicular to the electric oscillations, but it reflects them when rotated through a right angle, so that the oscillations take place along the conducting wires ; thus representing a kind of analyzer proving the existence of polarized light. The receiver itself also acted as analyzer, for if rotated much it failed to feel the disturbance.

Conducting sheets, even thin ones, were very opaque to the electrical radiation ; but non-conducting obstacles, even such as wood, interrupt it very little, and Dr. Hertz remarks, "not without wonder," that the door separating the room containing the source of radiation from that containing the detecting receiver might be shut without inter-

X

cepting the communication. The secondary sparks were still observed.

But the most crucial test yet applied is that of refraction. A great prism of pitch was made, its faces more than a yard square, and its refracting angle about 30°. This being interposed in the path of the electric rays, they were lost to the receiver until it was shifted considerably Adjusting it till its sparks were again at a maximum, it was found that the rays had been bent by the pitch prism, when set symmetrically, some 22° out of their original course, and hence that the pitch had an index of refraction for these 2-foot waves about 1·7.

190. These are great experiments. As I write, the latest of them are but a month or two old, and they are manifestly only a beginning. Most of the earlier ones are very simple, and have already been repeated.[1] They seem likely to settle many doubtful points. There has been a long-standing controversy in optics, nearly as old as the century, as to whether the direction of the vibrations was in, or was perpendicular to, the plane of polarization ; in other words, whether it was the elasticity or the density of the ether which varied in dense media ; or, in Maxwell's theory, whether

[1] See Fitzgerald and Trouton, *Nature*, Vol. 39, p. 391 ; also Dr. Dragoumis, *Nature*, Vol. 39, p. 548. Also Lodge and Howard, *Phil. Mag.* July 1889.

it was the electro-magnetic or the electrostatic disturbance that coincided with that plane. This point
has indeed by the exertion of extraordinary power
been almost settled already, through the consideration of common optical experiments ; but now that
we are able electrically to produce radiation with
a full knowledge of what we are doing, of its directions of vibration and all about it, the complete
solution of this and of many another recondite
optical problem may be expected during the next
decade to drop simply and easily into our hands.

We have now a real undulatory theory of light,
no longer based on analogy with sound, and its
inception and early development are among the
most tremendous of the many achievements of the
latter half of the nineteenth century.

In 1865, Maxwell stated his theory of light.
Before the close of 1888 it is utterly and completely
verified. Its full development is only a question of
time, and labour, and skill. The whole domain of
Optics is now annexed to Electricity, which has
thus become an imperial science.

APPENDED LECTURES.

(The following lectures bearing on the subject of this book are here conveniently appended. In one or two places the date of their delivery must be taken into account.)

LECTURE I.

THE RELATION BETWEEN ELECTRICITY AND LIGHT.[1]

EVER since the subject on which I have the honour to speak to you to-night was arranged, I have been astonished at my own audacity in proposing to deal in the course of sixty minutes with a subject so gigantic and so profound that a course of sixty lectures would be inadequate for its thorough and exhaustive treatment.

I must indeed confine myself carefully to some few of the typical and most salient points in the relation between electricity and light, and I must economize time by plunging at once into the middle of the matter without further preliminaries.

Now when a person is setting off to discuss the relation between electricity and light it is very natural and very proper to pull him up short with the two

[1] Delivered at the London Institution on December 16, 1880.

questions : What do you mean by electricity? and
What do you mean by light? These two questions
I intend to try briefly to answer. And here let me
observe that in answering these fundamental questions
I do not necessarily assume a fundamental ignorance
on your part of these two agents, but rather the
contrary ; and must beg you to remember that if
I repeat well-known and simple experiments before
you, it is for the purpose of directing attention to
their real meaning and significance, not to their obvious
and superficial characteristics : in the same way that
I might repeat the exceedingly familiar experiment
of dropping a stone to the earth if we were going to
define what we meant by gravitation.

Now then we will ask first, What is Electricity?
and the simple answer must be, We don't know. Well,
but this need not necessarily be depressing. If the
same question were asked about Matter, or about
Energy, we should have likewise to reply, No one
knows.

But then the term Matter is a very general one,
and so is the term Energy. They are heads, in fact,
under which we classify more special phenomena.

Thus if we were asked What is sulphur? or What
is selenium? we should at least be able to reply, A
form of matter ; and then proceed to describe its properties, *i.e.* how it affected our bodies and other bodies.

Again, to the question, What is heat? we can reply,

A form of energy; and proceed to describe the peculiarities which distinguish it from other forms of energy.

But to the question, What is electricity? we have no answer pat like this. We cannot assert that it is a form of matter, neither can we deny it; on the other hand, we certainly cannot assert that it is a form of energy, and I should be disposed to deny it. It may be that electricity is an entity *per se*, just as matter is an entity *per se*.

Nevertheless I can tell you what I mean by electricity by appealing to its known behaviour.

Here is a battery—that is, an electricity pump: it will drive electricity along. Prof. Ayrton is going, I am afraid, to tell you, on the 20th of January next, that it *produces* electricity; but if he does, I hope you will remember that that is exactly what neither it nor anything else can do. It is as impossible to generate electricity in the sense I am trying to give the word, as it is to produce matter. Of course I need hardly say that Prof. Ayrton knows this perfectly well; it is merely a question of words, *i.e.* of what you understand by the word electricity.[1]

---

[1] Or rather of what one understands by the word "produces." The title of Prof. Ayrton's lecture was "The Production of Electricity"; and it was to guard persons from supposing that it is right to speak of the generation or creation of electricity in the same way as it is possible to speak of the generation or creation (or, as it is often called, "production") of heat, that I gave this caution.

I want you then to regard this battery and all
electrical machines and batteries as kinds of electricity
pumps, which drive the electricity along through the
wire very much as a water-pump can drive water along
pipes, and that no electric machine can manufacture
electricity any more than a pump can manufacture
water. While the flow of electricity is going on, the
wire manifests a whole series of properties, which are
called the properties of the current.

[Here were shown an ignited platinum wire, the
electric arc between two carbons, an electric machine
spark, an induction-coil spark, and a vacuum tube
glow. Also a large nail was magnetized by being
wrapped in the current, and two helices were suspended and seen to direct and attract each other.]

To make a magnet, then, we only need a current of
electricity flowing round and round in a whirl. A
vortex or whirlpool of electricity is in fact a magnet ;
and *vice versâ*. And these whirls have the power of
directing and attracting other previously existing
whirls according to certain laws, called the laws of
magnetism. And, moreover, they have the power of
exciting fresh whirls in neighbouring conductors, and
of repelling them according to the laws of diamagnetism. The theory of the actions is known ;
though the nature of the whirls, as of the simple
stream of electricity, is at present unknown.

[Here was shown a large electro-magnet and an

induction-coil vacuum discharge spinning round and round when placed in its field (Fig. 24).]

So much for what happens when electricity is made to travel along conductors, *i.e.* when it travels along like a stream of water in a pipe, or spins round and round like a whirlpool.

But there is another set of phenomena, usually regarded as distinct and of another order, but which are not so distinct as they appear, which manifest themselves when you join the pump to a piece of glass or any non-conductor and try to force the electricity through that. You succeed in driving some through, but the flow is no longer like that of water in an open pipe ; it is as if the pipe were completely obstructed by a number of elastic partitions, or diaphragms. The water cannot move without straining and bending these diaphragms, and if you allow it, these strained partitions will recover themselves and drive the water back again. [Here was explained the process of charging a Leyden jar, and the model (Fig. 11) was shown.] The essential thing to remember is that we may have electrical energy in two forms, the static and the kinetic ; and it is therefore also possible to have the rapid alternation from one of these forms to the other, called vibration.

Now we will pass to the second question : What do you mean by light ? And the first and obvious answer is, Everybody knows. And everybody that is

not blind does know to a certain extent. We have a
special sense-organ for appreciating light, whereas we
have none for electricity. Nevertheless, we must admit that we really know very little about the intimate
nature of light—very little more than about electricity.
But we do know this, that light is a form of energy ;
and, moreover, that it is energy rapidly alternating
between the static and the kinetic forms—that it is, in
fact, a special kind of energy of vibration. We are absolutely certain that light is a periodic disturbance in
some medium, periodic both in space and time ; that is
to say, the same appearances regularly recur at certain
equal intervals of distance at the same time, and also
present themselves at equal intervals of time at the same
place ; that in fact it belongs to the class of motions
called by mathematicians undulatory or wave motions.

The wave motion in this model (Powell's wave apparatus) results from the simple up-and-down motion
popularly associated with the term *wave*. But when
a mathematician calls a thing a wave he means that
the disturbance is represented by a certain general
type of formula, not that it is an up-and-down motion,
or that it looks at all like those things on the top of
the sea. The motion of the surface of the sea falls
within that formula, and hence is a special variety of
wave motion, and the term wave has acquired in
popular use this signification and nothing else. So
that when one speaks ordinarily of a wave or undula-

tory motion one immediately thinks of something heaving up and down, or even perhaps of something breaking on the shore. But when we assert that the form of energy called light is *undulatory*, we by no means intend to assert that anything whatever is moving up and down, or that the motion, if we could see it, would be anything at all like what we are accustomed to in the ocean. The kind of motion is unknown ; we are not even sure that there is anything like motion in the ordinary sense of the word at all.

Now how much connection between electricity and light have we perceived in this glance into their natures ? Not much truly. It amounts to about this : That on the one hand electrical energy may exist in either of two forms—the static form, when insulators are electrically strained by having had electricity driven partially through them (as in the Leyden jar), which strain is a form of energy because of the tendency to discharge and do work ; and the kinetic form, where electricity is moving bodily along through conductors or whirling round and round inside them, which motion of electricity is a form of energy, because the conductors and whirls can attract or repel each other and thereby do work.

And, on the other hand, that light is the rapid alternation of energy from one form to another—from the static form where the medium is strained, to the kinetic form when it moves. It is just conceivable then that the static form of the energy of light is *electro*-static—that is, that the medium is *electrically* strained—and that the kinetic form of the energy of light is *electro*-kinetic—that is, that the motion is not ordinary motion, but electrical motion ; in fact that light is an electrical vibration, not a material one.

On November 5 last year there died at Cambridge a man in the full vigour of his faculties—such faculties as do not appear many times in a century—whose chief work has been the establishment of this very fact, the discovery of the link connecting light and electricity ; and the proof—for I believe it amounts to a proof—that they are different manifestations of one and the same class of phenomena : that light is, in fact, an electro-magnetic disturbance. The premature death of James Clerk Maxwell is a loss to science which appears at present utterly irreparable, for he was engaged in researches that no other man can hope as yet adequately to grasp and follow out ; but fortunately it did not occur till he had published his book on *Electricity and Magnetism*, one of those immortal productions which exalt one's idea of the mind of man, and which has been mentioned by competent critics in the same breath as the *Principia* itself.

But it is not perfect like the *Principia ;* much of it is rough-hewn, and requires to be thoroughly worked out. It contains numerous misprints and errata, and part of the second volume is so difficult as to be almost unintelligible. Some, in fact, consists of notes written for private use, and not prepared for publication. It seems next to impossible now to mature a work silently for twenty or thirty years, as was done by Newton two and a half centuries ago. But a second edition was preparing, and much might have been improved in form if life had been spared to the illustrious author.

The main proof of the electro-magnetic theory of light is this. The rate at which light travels has been measured many times, and is pretty well known. The rate at which an electro-magnetic wave disturbance would travel, if such could be generated, can be also determined by calculation from electrical measurements. The two velocities agree exactly. This is the great physical constant known as the ratio " v," which so many physicists have been measuring, and are likely to be measuring for some time to come (§ 138).

Many and brilliant as were Maxwell's discoveries, not only in electricity, but also in the theory of the nature of gases, and in molecular science generally, I cannot help thinking that if one of them is more striking and more full of future significance than the rest, it is the one I have just mentioned—the theory that light is an electrical phenomenon.

The first glimpse of this splendid generalization

was caught in 1845, five and thirty years ago, by that prince of pure experimentalists, Michael Faraday. His reasons for suspecting some connection between electricity and light are not clear to us—in fact they could not have been clear to him; but he seems to have felt a conviction that if he only tried long enough, and sent all kinds of rays of light in all possible directions across electric and magnetic fields in all sorts of media, he must ultimately hit upon something. Well, this is very nearly what he did. With a sublime patience and perseverance which remind one of the way Kepler hunted down guess after guess in a different field of research, Faraday combined electricity, or magnetism, and light in all manner of ways, and at last he was rewarded with a result. And a most out-of-the-way result it seemed. First you have to get a most powerful magnet and very strongly excite it; then you have to pierce its two poles with holes, in order that a beam of light may travel from one to the other along the lines of force; then, as ordinary light is no good, you must get a beam of plane-polarized light and send it between the poles. But still no result is obtained until, finally, you interpose a piece of a rare and out-of-the-way material which Faraday had himself discovered and made, a kind of glass which contains borate of lead, and which is very heavy, or dense, and which must be perfectly annealed.

And now, when all these arrangements are completed, what is seen is simply this, that if an analyzer is arranged to stop the light and make the field quite dark before the magnet is excited, then directly the battery is connected and the magnet called into action a faint and barely perceptible brightening of the field occurs ; which will disappear if the analyzer be slightly rotated. [The experiment was then shown.] Now no wonder that no one understood this result. Faraday himself did not understand it at all : he seems to have thought that the magnetic lines of force were rendered luminous, or that the light was magnetized ; in fact he was in a fog, and had no idea of its real significance. Nor had anyone. Continental philosophers experienced some difficulty and several failures before they were able to repeat the experiment. It was in fact discovered too soon, before the scientific world was ready to receive it, and it was reserved for Sir William Thomson briefly, but very clearly, to point out, and for Clerk Maxwell more fully to develop, its most important consequences.

[The principle of the experiment was then illustrated by the aid of a mechanical model. The model was a Wheatstone photometer consisting of one cogged circle rolling inside a fixed outer circle of twice the diameter, so that a bead attached to the inner one described some ellipse. An extra adjustment was provided whereby the bead could be set exactly over the circumference of the smaller wheel: it then describes a straight line, a diameter of the large circle, with a simple harmonic motion ; and this simple harmonic motion is actually compounded of two equal opposite circular motions, viz. the revolution of the centre of the smaller wheel, and the revolution of the bead about this moving centre in an opposite direction and at the same speed.

The whole instrument was mounted in such a way that it could be slowly rotated one way or other by a second handle and endless screw ; by this means one of these circular motions was accelerated and the other retarded, and as a consequence the path of the oscillating bead slowly rotated, describing a more complicated hypocycloid, and representing the rotation of the direction of vibration of light (§ 172).]

This is the fundamental experiment which probably suggested Clerk Maxwell's theory of light ; but of late years many fresh facts and relations between electricity and light have been discovered, and at the present time they are tumbling in in great numbers.

It was found by Faraday that many other transparent media besides heavy glass would show the phenomenon if placed between the poles : only in a less degree ; and the very important observation that air itself exhibits the same phenomenon, though to an exceedingly small extent, has just been made by Kundt and Röntgen in Germany.

Dr. Kerr, of Glasgow, has extended the result to opaque bodies, and has shown that if light be passed through magnetized *iron* its plane is rotated. The film of iron must be exceedingly thin, because of its opacity, and hence, though the intrinsic rotating power of iron is undoubtedly very great, the observed rotation is exceedingly small and difficult to observe ; and it is only by very remarkable patience and care and ingenuity that Dr. Kerr has obtained his result.   Mr. Fitzgerald, of Dublin, has examined the question mathematically, and has shown that Maxwell's theory would have enabled Dr. Kerr's result to be predicted.

Another requirement of the theory is that bodies which are transparent to light must be insulators or non-conductors of electricity, and that conductors of electricity are necessarily opaque to light.   Simple observation amply confirms this ; metals are the best conductors, and are the most opaque bodies known. Insulators such as glass and crystals are transparent whenever they are sufficiently homogeneous ; and the very remarkable researches of Prof. Graham Bell in the last few months have shown that even *ebonite*, one of the most opaque insulators to ordinary vision, is certainly transparent to some kinds of radiation, and transparent to no small degree.

[The reason why transparent bodies must insulate, and why conductors must be opaque, was here illustrated by mechanical models.

The model which represented a dielectric has already been depicted in Fig. 8 ; and when the cord threading all the elastically supported balls is vibrated, waves travel readily through it.

The model which represented a metallic conductor is shown here in Fig. 54. It has its wooden balls sliding on smooth brass rods so that they have no tendency to recoil to a settled position but remain where placed. On shaking the cord connecting these

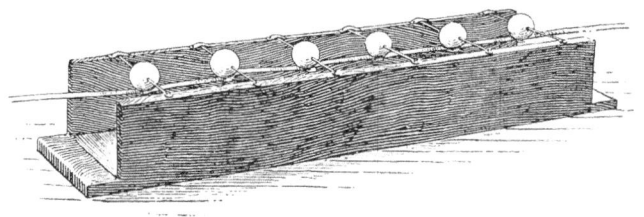

FIG. 54.—Rude model to pair with Fig. 8 (p. 43) and to call attention to some of the differences between a metal and an insulator.

balls the waves penetrate a certain small depth into the medium but fail to get through it.

The two models were connected in series, and waves which had been transmitted along the cord by one were partly quenched, partly reflected, by the other.]

A further consequence of the theory is that the velocity of light in a transparent medium will be affected by its electrical strain constant ; in other words, that its refractive index will bear some close but not yet quite ascertained relation to its specific

inductive capacity. Experiment has partially confirmed this, but the confirmation is as yet very
incomplete.

But there are a number of results not predicted by
theory, and whose connection with theory is not
clearly made out. We have the fact that light falling
on the platinum electrode of a voltameter generates a
current; first observed, I think, by Sir W. R. Grove—
at any rate it is mentioned in his *Correlation of
Forces*—extended by Becquerel and Robert Sabine
to other substances, and now being extended to
fluorescent and other bodies by Prof. Minchin. And
finally—for I must be brief—we have the remarkable
action of light on selenium. This fact was discovered
accidentally by an assistant in the laboratory of Mr.
Willoughby Smith, who noticed that a piece of
selenium conducted electricity very much better when
light was falling upon it than when it was in the dark.
The light of a candle is sufficient, and instantaneously
brings down the resistance to something like one-fifth
of its original value.

I could show you these effects, but there is not much
to see; it is an intensely interesting phenomenon, but
its external manifestation is not striking—any more
than Faraday's heavy glass experiment was.

This is the phenomenon which, as you know, has
been utilized by Prof. Graham Bell in that most
ingenious and striking invention, the photophone.

By the kindness of Prof. Silvanus Thompson I have a few slides to show the principle of the invention, and Mr. Shelford Bidwell has been good enough to lend me his home-made photophone, which answers exceedingly well for short distances.

I have now trespassed long enough upon your patience, but I must just allude to what may very likely be the next striking popular discovery, and that is the transmission of light by electricity ; I mean the transmission of such things as views and pictures by means of the electric wire.   It has not yet been done, but it seems already theoretically possible, and it may very soon be practically accomplished.

LECTURE II.

THE ETHER AND ITS FUNCTIONS.[1]

I HOPE that no one has been misled by an error in the printing of the title of this lecture, viz. the omission of the definite article before the word ether, into supposing that I am going to discourse on chemistry and the latest anæsthetic; you will have understood, I hope, that "ether" means *the* ether, and that the ether is the hypothetical medium which is supposed to fill otherwise empty space.

The idea of an ether is by no means a new one. As soon as a notion of the enormous extent of space had been grasped, by means of astronomical discoveries, the question presented itself to men's minds, what was in this space? was it full, or was it empty? and the question was differently answered by different metaphysicians. Some felt that a vacuum was so

---

[1] Delivered at the London Institution on December 28, 1882.

abhorrent a thing that it could not by any possibility exist anywhere, but that Nature would not be satisfied unless space were perfectly full. Others, again, felt that *empty* space could hardly exist; that it would shrink up to nothing like a pricked bladder unless it were kept distended by something material. In other words, they made matter the condition of extension. On the other hand, it was contended that, however objectionable the idea of empty space might be, yet emptiness was a necessity in order that bodies might have room to move ; that, in fact, if all space were perfectly full of matter everything would be jammed together, and nothing like free attraction or free motion of bodies round one another could go on.

And indeed there are not wanting philosophers at the present day who still believe something of this same kind, who are satisfied to think of matter as consisting of detached small particles acting on one another with forces varying as some inverse power of the distance, and who, if they can account for a phenomenon by an action exerted across empty space, are content to go no further, nor seek the cause and nature of the action more closely.[1]

Now metaphysical arguments, in so far as they

---

[1] In illustration of this statement an article has since appeared in the January number of the *Philosophical Magazine* for 1883, by Mr. Walter Browne.

have any weight or validity whatever, are unconscious appeals to experience ; a person endeavours to find out whether a certain condition of things is by him conceivable, and if it is not conceivable he has some *primâ facie* ground for asserting that it probably does not exist. I say he has *some* ground, but whether it be much or little depends partly on the nature of the thing thought of, whether it be fairly simple or highly complex, and partly on the range of the man's own mental development, whether his experience be wide or narrow.

If a highly-developed mind, or set of minds, find a doctrine about some comparatively simple and fundamental matter absolutely unthinkable, it is an evidence, and is accepted as good evidence, that the unthinkable state of things is one that has no existence ; the argument being that if it did exist, either it or something not wholly unlike it would have come within the range of experience. We have no further evidence than this for the statement that two straight lines cannot inclose a space, or that the three angles of a triangle are equal to two right angles.

Nevertheless there is nothing final about such an argument ; all that the inconceivability of a thing really proves, or can prove, is that nothing like it has ever come within the thinker's experience ; and this proves nothing as to the reality or non-reality of the thing, unless his experience of the same kind of things has been so extensive as to make it reasonably probable that if such a thing had existed it would not have been so completely overlooked.

The experience of a child or a dog, on ordinary scientific phenomena, therefore, is worth next to nothing ; and as the experience of a dog is to ordinary science, so is the experience of the human race to some higher phenomena, of which they at present know nothing, and against the existence of which it is perfectly futile and presumptuous to bring forward arguments about their being inconceivable, as if they were likely to be anything else.

Now if there is one thing with which the human race has been more conversant from time immemorial than another, and concerning which more experience has been unconsciously accumulated than about almost anything else that can be mentioned, it is *the action of one body on another ;* the exertion of force by one body upon another, the transfer of motion and energy from one body to another ; any kind of effect, no matter what, which can be produced in one body by means of another, whether the bodies be animate or inanimate.   The action of a man in felling a tree, in thrusting a spear, in drawing a bow ; the action of the bow again on the arrow, of powder on a bullet, of a horse on a cart ; and again, the action of the earth on the moon, or of a magnet on iron.   Every activity of every kind that we are conscious of may be taken as an illustration of the action of one body on another.

Now I wish to appeal to this mass of experience, and to ask, is not the direct action of one body on another across empty space, and with no means of communication whatever, is not this absolutely unthinkable? We must not answer the question off-hand, but must give it due consideration, and we shall find, I think, that wherever one body acts on another by obvious contact, we are satisfied and have a feeling that the phenomenon is simple and intelligible; but that whenever one body apparently acts on another at a distance, we are irresistibly impelled to look for the connecting medium.

If a marionette dances in obedience to a prompting hand above it, any intelligent child would feel for the wire, and if no wire or anything corresponding to it was discovered, would feel that there was something uncanny and magical about the whole thing. Ancient attempts at magic were indeed attempts to obtain results without the trouble of properly causing them, to build palaces by rubbing rings or lanterns, to remove mountains by a wish instead of with the spade and pickaxe, and generally to act on bodies without any real means of communication; and modern disbelief in magic is simply a statement of the conviction of mankind that all attempts in this direction have turned out failures, and that action at a distance is impossible.

If a man explained the action of a horse or a cart by saying that there was an attraction between them varying as some high direct power of the distance, he would not be saying other than the truth—the facts may be so expressed—but he would be felt to be giving a wretchedly lame explanation, and anyone who simply pointed out the traces would be going much more to the root of the matter. Similarly with the attraction of a magnet for a distant magnetic pole. To say that there is an attraction as the inverse cube of the distance between them is true, but it is not the whole truth ; and we should be obliged to anyone who will point out the traces, for traces we feel sure there are.

If anyone tries to picture clearly to himself the action of one body on another without any medium of communication whatever, he must fail. A medium is instinctively looked for in most cases ; and if not in all, as in falling weights or magnetic attraction, it is only because custom has made us stupidly callous to the real nature of these forces.

When we see a vehicle bowling down-hill without any visible propelling force, we ought to regard it with the same mixture of curiosity and wonder as the Chinaman felt when he saw for the first time in the streets of Chicago a tram-car driven by a rope buried in a pipe underground. The attachment to these cars comes through a narrow slit in the pipe, and is quite unobtrusive. After regarding the car with open-mouthed astonishment for some time, the Chinaman made use of the following memorable exclamation, "No pushee—No pullee—Go like mad!" He was a philosophic Chinaman.

Remember, then, that whenever we see a thing being moved we must look for the rope; it may be visible or it may be invisible, but unless there is either "pushee" or "pullee" there can be no action. And if you further consider a pull it resolves itself into a push; to pull a thing towards you, you have to put your finger behind it and push; a horse is said to pull a cart, but he is really pushing at the collar; an engine pushes a truck by means of a hook and eye; and so on.

There is still the further very important and difficult question as to why the parts hang together, and why when you push one part the rest follows. Cohesion is a very striking fact, and an explanation of it is much to be desired; I shall have a little more to say about it later, but at present we have nothing more than an indication of the direction in which an explanation seems possible. We cannot speak distinctly about those actions which are as yet mysterious to us; but concerning those which are comparatively simple and intelligible we may make this general statement :—The only way of acting on a body directly is to push it behind.

There must be contact between bodies before they can directly act on each other ; and if they are not in contact with each other and yet act, they must both be in contact with some third body which is the medium of communication, the rope.

Consider now for an instant the most complex case, the action of one animate body on another not touching it. To call the attention of a dog, for instance, there are several methods : one plan is to prod him with a stick, another is to heave a stone at him, a third is to whistle or call, while a fourth is to beckon him by gesture, or, what is essentially the same process, to flash sunlight into his eye with a mirror. In the first two of these methods the media of communication are perfectly obvious—the stick and the stone ; in the third, the whistle, the medium is not so obvious, and this case might easily seem to a savage like action at a distance, but we know of course that it is the air, and that if the air between be taken away, all communication by sound is interrupted. But the fourth or optical method is not so interrupted ; the dog can see through a vacuum perfectly well, though he cannot hear through it ; but what the medium now is which conveys the impression is not so well known. The sun's light is conveyed to the earth by such a medium as this across the emptiness of planetary space.

The only remaining typical plans of acting on the dog would be either by electric or magnetic attractions, or by mesmerism, and I would have you seek for the medium which conveys these impressions with just as great a certainty that there is one as you feel in any of the other cases.

Leaving these more mysterious and subtle modes of communication, let us return to the two most simple ones, viz. the stick and the stone. *These two are representative of the only possible fundamental modes of communication between distant bodies,* for one is compelled to believe that every more occult mode of action will ultimately resolve itself into one or other of these two. The stick represents the method of communication by continuous substance ; the stone represents the communication by actual transfer of matter, or, as I shall call it, the projectile method. There are no other known methods for one body to act on another than by these two—by continuous medium, and by projectile.

We know one clear and well-established example of the projectile method, viz. the transmission of pressure by gases. A gas consists of particles perfectly independent of each other, and the only way in which they can act on each other is by blows. The pressure of the air is a bombardment of particles, and actions are transmitted through gases as through a row of ivory balls. Sound is propagated by each particle receiving a knock and passing it on to the next, the final effect being much the same as if the first struck particles had been shot off through the whole distance.

The explanation of the whole behaviour of gases in this manner is so simple and satisfactory, and moreover is so certainly the true account of the matter, that we are naturally tempted to ask whether this projectile theory is not the key to the universe, and whether every kind of action whatever cannot be worked out on this hypothesis of atoms blindly driving about in all directions at perfect random, and with complete independence of each other except when they collide.[1] And accordingly we have the corpuscular theories of light and of gravitation : both account for their respective phenomena by a battering of particles. The corpuscular theory of gravitation is, however, full of difficulties, for it is not obvious according to it why the weight of a plate is the same when held edgeways as when held broadside on, in the stream of corpuscles ; while it is surprising (as indeed it perhaps is on any hypothesis) that the weight of a body is the same in the solid, liquid, and gaseous states. It has been attempted to explain cohesion also on the same hypothesis, but the diffi-

---

[1] To this hypothesis Mr. Tolver Preston has addressed himself with much ingenuity,

culties, which were great enough before, are now enormous ; and to me at any rate it seems that it is only by violent straining and by improbable hypotheses that we can explain all the actions of the universe by a mere battery of particles.

Moreover, it is difficult to understand what the atoms themselves can be like, or how they can strike and bound off one another without yielding to compression and then springing out again like two elastic balls ; it is difficult to understand the elasticity of really ultimate hard particles. And if the atoms are not such hard particles, but are elastic and yielding, and rebound from one another according to the same sort of law that ivory balls do : of what are they composed ? We shall have to begin all over again, and explain the cohesion and elasticity of the parts of the atom.

The more we think over the matter, the more are we compelled to abandon mere impact as a complete explanation of action in general. But if this be so we are driven back upon the other hypothesis, the only other, viz. communication by continuous medium.

We must begin to imagine a continuous connecting medium between the particles—a substance in which they are embedded, which penetrates into all their interstices, and extends without break to the remotest limits of space. Once grant this, and difficulties begin rapidly to disappear. There is now continuous

Z

contact between the particles of bodies, and if one is pushed the others naturally receive the motion. The atoms of gas are impinging as before, but we have now a different idea of what impact means.

Gravitation is explainable by differences of pressure in the medium, caused by some action between it and matter not yet understood. (See page 352.) Cohesion is explainable also probably in the same way.

Light consists of undulation or waves in the medium; while electricity is turning out quite possibly to be an aspect of a part of the very medium itself.

The medium is now accepted as a necessity by all modern physicists, for without it we are groping in the dark, with it we feel we have a clue which, if followed up, may lead us into the innermost secrets of Nature. It has as yet been followed up very partially, but I will try and indicate the directions in which modern science is tending.

The name you choose to give to the medium is a matter of very small importance, but " the ether " is as good a name for it as another.

As far as we know it appears to be a perfectly homogeneous incompressible continuous body, incapable of being resolved into simpler elements or atoms; it is, in fact, continuous, not molecular. There is no other body of which we can say this, and hence the properties of ether must be somewhat different from those of ordinary matter. But there is little difficulty in picturing a continuous substance to ourselves, inasmuch as the molecular and porous nature of ordinary matter is by no means evident to the senses, but is an inference of some difficulty.

Ether is often called a fluid, or a liquid, and again it has been called a solid and has been likened to a jelly because of its rigidity; but none of these names are very much good ; all these are molecular groupings and therefore not like ether ; let us think simply and solely of a continuous frictionless medium possessing inertia, and the vagueness of the notion will be nothing more than is proper in the present state of our knowledge.

We have now to try and realize the idea of a perfectly continuous, subtle, incompressible substance pervading all space and penetrating between the molecules of all ordinary matter, which are embedded in it, and connected with one another by its means. And we must regard it as the one universal medium by which all actions between bodies are carried on. This, then, is its function—to act as the transmitter of motion and of energy.

First consider the propagation of light.

Sound is propagated by direct excursion and impact of the atoms of ordinary matter. Light is not so propagated. How do we know this ?

(1) Because of its speed, 3×10^{10} centimetres per

second, which is greater than anything transmissible by ordinary matter.

(2) Because of the kind of vibration, as revealed by the phenomena of polarization.

The vibrations of light are not such as can be transmitted by a set of disconnected molecules ; if by molecules at all, it must be by molecules connected into a solid, *i.e.* by a body with rigidity. Rigidity means active resistance to shearing stress, *i.e.* to alteration in shape ; it is also called *elasticity of figure;* it is by the possession of rigidity that a solid differs from a fluid. For a body to transmit vibrations at all it must possess inertia ; transverse vibrations can only be transmitted by a body with rigidity. All matter possesses inertia, but fluids only possess volume elasticity, and accordingly can only transmit longitudinal vibrations. Light consists of transverse vibrations ; air and water have no rigidity, yet they are transparent, *i.e.* transmit transverse vibrations ; hence it must be the ether inside them which really conveys the motion, and the ether must have properties which, if it were ordinary matter, we should style *inertia* and *rigidity*. No highly rarefied air will serve the purpose ; the ether must be a distinct body. Air may *exist* indeed in planetary space, even to infinity, but if so it is of almost infinitesimal density compared with the ether there. It is easy to calculate the density of the atmosphere

at any height above the earth's surface, supposing
other bodies absent and supposing the temperature
constant. (All numbers following are in C.G.S. units.)

The density of the air at a distance of n earth
radii from the centre of the earth is equal to a
quarter the density here divided by $10^{350\frac{n-1}{n}}$. So at a
height of only 4000 miles above the surface, the
atmospheric density is a number with 127 ciphers
after the decimal point before the significant figures
begin.[1] The density of ether, on the other hand,
has been calculated by Sir William Thomson from
data furnished by Pouillet's experiments on the
energy of sunlight, and from a justifiable guess as
to the amplitude of a vibration ; and it comes out
about 10^{-18}, a number with only 17 ciphers before
the significant figures. In inter-planetary space,
therefore, all the air that exists is utterly negligible ;
the density of the ether there, though small, is
enormous by comparison. [See also page 235.]

Once given the density of the ether, its rigidity
follows at once, because the ratio of the rigidity
to the density is the square of the velocity of trans-

[1] I have left this statement in, because it is a view which has been
apparently held by high authority that the atmosphere has no limit.
To me I confess it appears much more reasonable to suppose that at a
certain height, which on the hypothesis of thorough stirring or convection equilibrum is only 16 or 17 miles, but is probably a good deal
more in reality (because rare air is very viscous), a free surface exists
although of very small density.

verse wave propagation, viz. in the case of ether, 9×10^{20}. The rigidity of ether comes out, therefore, to be about 900. The most rigid solid we know is steel, and compared with its rigidity, viz. 8×10^{11}, that of ether is insignificant. Neither steel nor glass, however, could transmit vibrations with anything like the speed of light, because of their great density. The rate at which transverse vibrations are propagated by crown glass is half a million centimetres per second— a considerable speed, no doubt, but the ether inside the glass transmits them 40,000 times as quick, viz. at twenty thousand million centimetres per second.

The ether outside the glass can do still better than this, it comes up to thirty thousand million, and the question arises what is the matter with the ether inside the glass that it can only transmit undulations at two-thirds the normal speed. Is it denser than free ether, or is it less rigid? Well, it is not easy to say; but the fact is certain that ether is somehow affected by the immediate neighbourhood of gross matter, and it appears to be concentrated inside it to an extent depending on the density of the matter. Fresnel's hypothesis is that the ether is really denser inside gross matter, that there is a sort of attraction between ether and the molecules of matter which results in an agglomeration or binding of some ether round each atom, and that this additional or bound ether belongs to the matter, and travels about with it. The *rigidity* of the bound ether Fresnel supposes to be the same as that of the free, except in some crystals.

If anything like this can be imagined, a measure of the relative density of the bound ether is easily given. For the inverse velocity-ratio of light is n (the index of refraction), and the density is inversely as the square of the velocity ; hence the density-measure is n^2. The density of ether in free space being called I, that inside matter has a density n^2, and the density of the bound portion of this is $n^2 - 1$.

This may all sound very fanciful, but something like it is sober truth ; not as it is here stated very likely, but the fact that $\left(1 - \dfrac{1}{n^2} \right)$th of the whole ether inside matter is bound to it and travels with it, while the remaining $\dfrac{1}{n^2}$th is free and blows freely through the pores, is fairly well established and confirmed by direct experiment (§ 118).

Consider the effect of wind on sound. Sound is travelling through the air at a certain definite rate depending simply on the average speed of the atoms in their excursions, and at the rate at which they therefore pass the knocks on ; if there is a wind carrying all the atoms bodily in one direction, naturally the sound will travel quicker in that direction than in the opposite. Sound travels quicker with the wind than against it. Now is it the same with light ?

does it too travel quicker with the wind ? Well that altogether depends on whether the ether is blowing along as well as the air ; if it is, then its motion must help the light on a little ; but if the ether is at rest, no motion of air or matter of any kind can make any difference. But according to Fresnel's hypothesis it is not wholly at rest nor wholly in motion ; the free is at rest, the bound is in motion ; and therefore the speed of light with the wind should be increased by an addition of $\left(1 - \dfrac{1}{n^2}\right)$th of the velocity of the wind. Utterly infinitesimal, of course, in the case of air, whose n is but a trifle greater than 1 ; but for water the fraction is 7-16ths, and Fizeau thought this not quite hopeless to look for. He accordingly devised a beautiful experiment, executed it successfully, and proved that when light travels with a stream of water, 7-16ths of the velocity of the water must be added to the velocity of the light ; and when it travels against the stream the same quantity must be subtracted, to get the true resultant velocity with which the light is travelling through space.

Arago suggested another experiment. When light passes through a prism, it is bent out of its course by reason of its diminished velocity inside the glass, and the refraction is strictly dependent on the re-tardation ; now suppose a prism carried rapidly forward through space, say at the rate of nineteen miles a second by the earth in its orbit, which is the quickest accessible carriage ; if the ether is all streaming freely through the glass, light passing through the prism will be less retarded when going with the ether than when going against it, and hence the bending will be different.

Maxwell tried the experiment in a very perfect form, but found no difference. If all the ether were free there would have been a difference ; if all the ether were bound to the glass there would have been a difference the other way ; but according to Fresnel's hypothesis there should be no difference, because according to it, the free ether, which is the portion in relative motion, has nothing to do with the refraction, it is the addition of the bound ether which causes the refraction, and this part is stationary relatively to the glass, and is not streaming through it at all.   Hence the refraction is the same whether the prism be at rest or in motion through space.[1]

An atom embedded in ether is vibrating and sending out waves in all directions ; the length of the wave depends on the period of the vibration, and

---

[1] Several of this class of experiments have been recently performed with consummate skill and with refined appliances by Mr. Michelson in America.   The result of his repetition of the Fizeau experiment is entirely confirmatory of Fizeau's result and of Fresnel's theory.   The results of some of the other experiments, having reference to the theory of aberration and the motion of the ether near the earth, are more puzzling, and seem discordant with ordinarily received notions at present.

different lengths of wave produce the different colour sensations. Now through free ether all kinds of waves appear to travel at the same rate ; not so through bound ether ; inside matter the short waves are more retarded than the long, and hence the different sizes of waves can be sorted out by a prism. Now a free atom has its own definite period of vibration, like a tuning-fork has, and accordingly sends out light of a certain definite colour or of a few definite colours, just as a tuning-fork emits sound of a certain definite pitch or of a few definite pitches called harmonics. By the pitch of the sound it is easy to calculate the rate of vibration of the fork ; by the colour of the light one can determine the rate of vibration of the atom.

When we speak of the atoms vibrating, we do not mean that they are wagging to and fro as a whole ; it is more likely that they are crimping themselves, that they are vibrating as a tuning-fork or a bell vibrates ; we know this because it is easy to make the free atoms of a gas vibrate. It is only in the gaseous state, indeed, that we can study the rate of vibration of an atom ; when they are packed closely together in a solid or liquid, they are cramped, and all manner of secondary vibrations are induced. They then, no doubt, wag to and fro also ; and in fact these con-strained vibrations are executed in every variety, but the simple periodicity of the free atom is lost.

To study the free atoms we take a gas—the rarer

the better: heat it, and then sort out the waves it produces in the ether by putting a triangular prism of bound ether in their path.

Why the bound ether retards different waves differently, or "disperses" the light, is quite unknown, beyond the fact that it has something to do with the size of the atoms of matter being comparable to the size of waves : being most nearly comparable to the smallest waves, and therefore affecting them most. It is not easy accurately to explain refraction, but it is extremely difficult to explain dispersion. However, the fact is undoubted, and more light will doubtless soon fall upon its theory.

The result of the prismatic analysis is to prove that every atom of matter has its own definite rate of vibration, as a bell has ; it may emit several colours or only one, and the number it emits may depend upon how much it is struck (or heated) ; but those it can emit are a perfectly definite selection, and depend in no way on the previous history of the atom. Every free atom of sodium, for instance, vibrates in the same way, and has always vibrated in the same way, whatever other element it may have been at intervals combined with, and whether it exists in the sun or in the earth, or in the most distant star. The same is true of every other kind of matter, each has its own mode of vibration which nothing but bondage changes ; and hence

has arisen a new chemical analysis, wherein substances are detected simply by observing the rate
of vibration of their free atoms, a branch of physical
chemistry called spectrum analysis.

The atoms are small bodies, and accordingly
vibrate with inconceivable rapidity.

An atom of sodium vibrates 5×10^{14} times in a
second; that is, it executes five hundred million
complete vibrations in the millionth part of a second.

This is about a medium pace, and the waves it emits
produce in the eye the sensation of a deep yellow.

4×10^{14} corresponds to red light, 7×10^{14} to blue.

An atom of hydrogen has three different periods,
viz. 4·577, 6·179, and 6·973, each multiplied by the
inevitable 10^{14}.

Atoms may, indeed, vibrate more slowly than
this, but the retina is not constructed so as to be
sensible of slower vibrations; however, thanks to
Capt. Abney, there are ways now of photographing
the effect of much slower vibrations, and thus of
making them indirectly visible; so we can now
hope to observe the motion of atoms over a much
greater range than the purely optical ones, and so
learn much more about them.[1]

---

[1] Still more perhaps may we now hope from the modified line
thermopile or Siemens pyrometer, which Prof. Langley has so ably
developed and used in a series of fine researches: the instrument
which he calls the "bolometer." Or from Mr. Boys's still more recent
" Radio-micrometer,"

The distinction between free and bound ether is forced on our notice by other phenomena than those of light. When we come to electricity, we find that some kind of matter has more electricity associated with it than others, so that for a given electromotive force we get a greater electric displacement; that the electricity is, as it were, denser in some kinds of matter than in others. The density of electricity in space being called I, that inside matter is called K, its specific inductive capacity. In optics the relative density of the ether inside matter was n^2, the square of the index of refraction (p. 343). These numbers appear to be the same.

Is the ether electricity then? I do not say so, neither do I think that in that coarse statement lies the truth; but that they are connected there can be no doubt.

What I have to suggest is that positive and negative electricity together may make up the ether, or that the ether may be sheared by electromotive forces into positive and negative electricity. Transverse vibrations are carried on by shearing forces acting in matter which resists them, or which possesses rigidity. The bound ether inside a conductor has no rigidity; it cannot resist shear; such a body is opaque. Transparent bodies are those whose bound ether, when sheared, resists and springs back again; such bodies are dielectrics.

We have no direct way of exerting force upon ether at all ; we can, however, act on it in a very indirect manner, for we have learnt how to arrange matter so as to cause it to exert the required shearing (or electromotive) force upon the ether associated with it. Continuous shearing force applied to the ether in metals produces a continuous and barely resisted stream of the two electricities in opposite directions : or a conduction current.

Continuous shearing force applied to the ether in transparent bodies produces an electric displacement accompanied by elastic resilience, and thus all the phenomena of electric induction (Chap. III.).

Some chemical compounds, consisting of binary molecules, *distribute* the bound ether of the molecule, at any rate as soon as it is split up by dissociation ; and, instead of each nascent radicle or atom taking with it neutral ether, one takes a certain definite quantity of positive, the other the same amount of negative, electricity. In the liquid state the atoms are capable of locomotion ; and a continuous shearing force applied to the ether in such liquids causes a continual procession of the matter and associated electricity, the positive one way, and the negative the other, and thus all the phenomena of electrolysis (Chap. IV.).

What I say about electricity, however, is not to be taken without salt ; you will not regard it as recognized truth, but as a tentative belief of your lecturer's which may be found to be more or less, and possibly more rather than less, out of accordance with facts. I can only say that it hangs phenomena together, and that it has been forced upon my belief in various ways.

Now what about the free ether of space, is it a conductor of electricity? There are certain facts which suggest that it is, and Edlund has suggested that it is an almost perfect conductor. When a sun-spot or other disturbance breaks out on the sun, accompanied as it is, no doubt, by violent electric storms, the electric condition of the earth is affected, and we have auroræ and magnetic disturbances. Is this by induction through space? or can it be due to conduction and the arrival of some microscopic portion of a derived current travelling our way?

For my part I cannot think the ether a conductor. Maxwell has shown that conductors must be opaque, and ether is nothing if not transparent ; one is driven, then, to conclude that what we call conduction does not go on except in the presence of ordinary matter—in other words, perhaps, that it is a phenomenon more connected with bound ether than with free.

But now, looking back to Fresnel's hypothesis of the extra density of ether inside gross matter, and also to the fact that it must be regarded as incompressible, the question naturally arises, How can it be densified by matter or anything else? Perhaps it is not ; perhaps matter only strains the ether towards itself, thus slackening its tension, as it were, inside bodies, not producing any real increase of density ; and this is roughly McCullagh's form of the undulatory theory. In this form gravitation may be held to be partially explained ; for two bodies straining at the ether in this way will tend to pull themselves together. Newton himself dimly suggested, in one of the queries appended to the later editions of his " Opticks," that gravitation would be produced if only matter exerted a kind of pressure on an all pervading ether, the pressure varying as the inverse distance. (See Appendix.)

He did not follow the idea up, however, because he had then no other facts to confirm him in his impression of the existence of such an ether, or to inform him concerning its properties. We now not only feel sure that an ether exists, but we know something of its properties ; and we also have learnt from light and from electricity, that some such action between matter and ether actually occurs, though how or why it occurs we do not yet know. I am therefore compelled to believe that this is certainly the direction in which an ultimate explanation of gravitation and of cohesion is to be looked for.

In thinking over the Fresnel and McCullagh forms of the undulatory theory, with a view to the reconciliation between them which appears necessary and imminent, one naturally asks, is there any such clear distinction to be drawn between ether and matter as we have hitherto tacitly assumed ? may they not be different modifications, or even manifestations, of the same thing ?

Again, when we speak of atoms vibrating, how can they vibrate ? of what are their parts composed ?

And now we come to one of the most remarkable and suggestive speculations of modern times—a speculation based on this experimental fact, that the elasticity of a solid may be accounted for by the motion of a fluid ; that a fluid in motion may possess rigidity.

I said that rigidity was precisely what no fluid possessed : at rest this is true ; in motion it is not true (§ 156).

Consider a perfectly flexible india-rubber O-shaped tube full of water ; nothing is more flaccid and limp. But set the water rapidly circulating, and it becomes at once stiff ; it will stand on end for a time without support; kinks in it take force to make, and are more or less permanent. A practicable form of this experiment is the well-known one of a flexible chain over a pulley, which becomes stiff as soon as it is set in rapid motion (page 166, footnote).

This is called a vortex stream-line, and a vortex is a thing built up of a number of such stream-lines. If they are arranged parallel to one another about a straight

A A

axis or core, we have a vortex cylinder, such as is easily produced by stirring a vessel of water or by pulling the plug out of a wash-hand basin ; or such as are made in the air on a large scale in America, and telegraphed over here, when they are called " cyclones," or " depressions." The *depression* is visible enough in the middle of revolving water. These vortices are wonderfully permanent things, and last a long time, though they sometimes break up unexpectedly.

Vortices need not have straight cores : they may have cores of various ring forms, the simplest being a circle. To make a vortex ring, we must take a plane disk of the fluid, and at a certain instant give to every atom in the disk a certain velocity forward, graduating the velocity according to its distance from the edge of the disk. We have as yet no means of doing this in a frictionless fluid, but with a fluid such as air and water it happens to be easy ; we have only to knock a little of the fluid suddenly out of a box through a sharp-edged hole, and the friction of the edges of the hole does what we want. The central portion travels rapidly forward, and returns round outside the core, rolling back towards the hole. But the impetus sends the whole forward, and none really returns ; it rolls on its outer circumference as a wheel rolls along a road. In a perfect fluid under conceivable circumstances it need not so roll forward, as there would be no friction, but in air or water a vortex ring has always a definite forward velocity, just as a locomotive driving-wheel has when it does not slip on the rails.

We have in these rings a real mass of air moving bodily forward, and it impinges on a face or a gas flame with some force. One is thus easily able to blow out a distant gas flame ten or twelve yards away by an invisible projectile of air. It is differentiated from the rest of the atmosphere by reason of its peculiar rotational motion. The ring may be rendered visible by means of smoke, but it is in no way improved by the addition except in the matter of visibility.

The cores of these rings are elastic—they possess rigidity ; the circular is their stable form, and if this is altered, they oscillate about it. Thus when two vortex rings impinge or even approach fairly near one another, they visibly deflect each other, and also cause each other to vibrate.

The theory of the impact or interference of vortex rings whose paths cross but which do not come very near together, has been quite recently worked out by Mr. J. J. Thomson. It is quite possible to make the rings vibrate without any impact, by serrating the opening out of which they are knocked. The simplest serration of a circle turns it into an ellipse, and here you have an elliptic ring oscillating from a tall to a squat ellipse and back again. Here is a four-waved opening, and the vibrations are by this very well shown. A six-waved opening makes the vibrations almost too small to be perceived at a distance, but still they are sometimes distinct.

The rings vibrate very much like a bell vibrates : perhaps very much like an atom vibrates. They have rigidity, although composed of fluid : they are composed of fluid in motion. These vortices are imperfect, they increase in size, and decrease in energy ; in a perfect fluid they would not do this, they would then be permanent and indestructible, but then also you would not be able to make them.

Now does not the idea strike you that atoms of matter may be vortices like these—vortices in a perfect fluid, vortices in the ether. This is Sir William Thomson's theory of matter. It is not yet proved to be true, but is it not highly beautiful ? a theory about which one may almost dare to say that it deserves to be true ? The atoms of matter, according to it, are not so much foreign particles imbedded in the all-pervading ether, as portions of it differentiated off from the rest by reason of their vortex motion, thus becoming virtually solid particles, yet with no transition of substance ; atoms indestructible and not able to be manufactured, not mere hard rigid specks, but each composed of whirling ether ; elastic, capable of definite vibration, of free movement, of collision. The crispations or crimpings of these rings illustrate the kind of way in which we may suppose an atom to vibrate. They appear to have all the properties of atoms except one, viz. gravitation ; and before the theory can be accepted, I think it must account for gravitation. This fundamental property of matter cannot be left over to be explained by an artificial battery of ultra-mundane corpuscles. We cannot go back to mere impact of hard bodies after having allowed ourselves a continuous medium. Vortex atoms must be shown to gravitate.

But then remember how small a force gravitation is. Ask any educated man whether two pound-masses of lead attract each other, and he will reply no. He is wrong, of course, but the force is exceedingly small. Yet it is the aggregate attraction of trillions upon trillions of atoms ; the *slightest* effect of each upon the ether would be sufficient to account for gravitation ; and no one can say that vortices do not exert some such residual, but uniform, effect on the fluid in which they exist, till second, third, and every other order of small quantities have been taken into account, and the theory of vortices in a perfect fluid worked out with the most final accuracy.

At present, however, the Thomsonian theory of matter is not a verified one ; it is, perhaps, little more than a speculation, but it is one that it is well worth knowing about, working at, and inquiring into. It may stand or it may fall ; but if it is the case, as I believe it is, that our notions of natural phenomena, though they often fall short, yet never exceed in grandeur the real truth of things, how splendid must be the real nature of matter if the Thomsonian hypothesis turns out to be inadequate and untrue.

I have now endeavoured to introduce you to the simplest conception of the material universe which has yet occurred to man—the conception, that is, of one universal substance, perfectly homogeneous and continuous and simple in structure, extending to the furthest limits of space of which we have any knowledge, existing equally everywhere ; some portions either at rest or in simple irrotational motion transmitting the undulations which we call light ; other portions in rotational motion, in vortices that is, and differentiated permanently from the rest of the medium by reason of this motion.

These whirling portions constitute what we call matter ; their motion gives them rigidity, and of them our bodies and all other material bodies with which we are acquainted are built up.

One continuous substance filling all space : which can vibrate as light ; which can be sheared into positive and negative electricity ; which in whirls constitutes matter ; and which transmits by continuity, and not by impact, every action and reaction of which matter is capable. This is the modern view of the Ether and its functions.

LECTURE III.

THE DISCHARGE OF A LEYDEN JAR.[1]

IT is one of the great generalizations established by Faraday, that all electrical charge and discharge is essentially the charge and discharge of a Leyden jar. It is impossible to charge one body alone. Whenever a body is charged positively, some other body is *ipso facto* charged negatively, and the two equal opposite charges are connected by lines of induction. The charges are, in fact, simply the ends of these lines, and it is as impossible to have one charge without its correlative as it is to have one end of a piece of string without there being somewhere, hidden it may be, split up into strands it may be, but somewhere existent, the other end of that string.

This I suppose familiar fact that all charge is virtually that of a Leyden jar being premised, our subject for this evening is at once seen to be a very

[1] Delivered at the Royal Institution of Great Britain, on Friday evening, March 8, 1889.

wide one, ranging in fact over the whole domain of electricity. For the charge of a Leyden jar includes virtually the domain of electrostatics ; while the discharge of a jar, since it constitutes a current, covers the ground of current electricity all except that portion which deals with phenomena peculiar to steady currents. And since a current of electricity necessarily magnetizes the space around it, whether it flow in a straight or in a curved path, whether it flow through wire or burst through air, the territory of magnetism is likewise invaded ; and inasmuch as a Leyden jar discharge is oscillatory, and we now know the vibratory motion called light to be really an oscillating electric current, the domain of optics is seriously encroached upon.

But though the subject I have chosen would permit this wide range, and though it is highly desirable to keep before our minds the wide-reaching import of the most simple-seeming fact in connection with such a subject, yet to-night I do not intend to avail myself of any such latitude, but to keep as closely and distinctly as possible to the Leyden jar in its homely and well-known form, as constructed out of a glass bottle, two sheets of tinfoil, and some stickphast.

The act of charging such a jar I have permitted myself now for some time to illustrate by the mechanical analogy of an inextensible endless cord able to circulate over pulleys, and threading in some portion of its length a row of tightly-gripping beads which are connected to fixed beams by elastic threads.

The cord is to represent electricity; the beads represent successive strata in the thickness of the glass of the jar, or, if you like, atoms of dielectric or

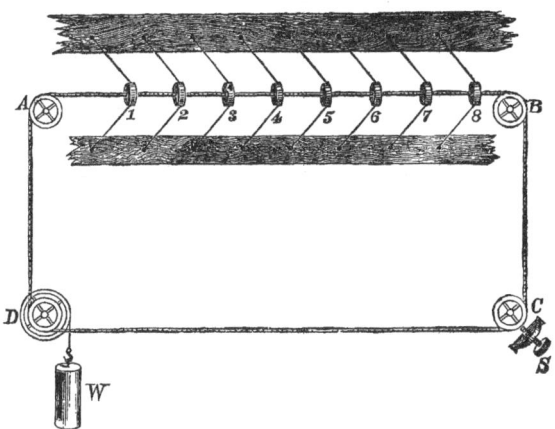

Fig. 55.—Mechanical analogy of a circuit partly *dielectric*; for instance, of a charged condenser. A is its positive coat, B its negative.

insulating matter. Extra tension in the cord represents negative potential, while a less tension (the nearest analogue to pressure adapted to the circumstances) represents positive potential. Forces applied to move the cord, such as winches or weights, are electromotive forces; a clamp or fixed obstruction represents a rheostat or contact-breaker; and an excess or defect of cord between two strata of matter represents a positive or a negative charge.

The act of charging a jar is now quite easily depicted as shown in the diagram.

To discharge the jar one must remove the charging E.M.F. and unclamp the screw, *i.e.* close the circuit. The stress in the elastic threads will then rapidly drive the cord back, the inertia of the beads will cause it to overshoot the mark, and for an instant the jar will possess an inverse charge. Back again the cord swings, however, and a charge of same sign as at first, but of rather less magnitude, would be found in the jar if the operation were now suspended. If it be allowed to go on, the oscillations gradually subside, and in a short time everything is quiescent, and the jar is completed discharged.

All this occurs in the Leyden jar, and the whole series of oscillations, accompanied by periodic reversal and re-reversal of the charges of the jar, is all accomplished in the incredibly short space of time occupied by a spark.

Consider now what the rate of oscillation depends on. Manifestly on the elasticity of the threads and on the inertia of the matter which is moved. Take the simplest mechanical analogy, that of the vibration of a loaded spring, like the reeds in a musical box. The stiffer the spring and the less the load, the faster it vibrates. Give a mathematician these data, and he will calculate for you the time the spring takes to execute one complete vibration, the "period" of its swing. [Loaded lath in vice.]

The electrical problem and the electrical solution are precisely the same. That which corresponds to the flexibility of the spring is in electrical language called static capacity, or, by Mr. Heaviside, permittance. That which corresponds to the inertia of ordinary matter is electro-magnetic inertia, or self-induction, or by Mr. Heaviside, inductance.

Increase either of these, and the rate of oscillation is diminished. Increasing the static capacity corresponds to lengthening the spring; increasing the self-induction corresponds to loading it.

Now the static capacity is increased simply by using a larger jar, or by combining a number of jars into a battery in the very old-established way. Increase in the self-induction is attained by giving the discharge more space to magnetize, or by making it magnetize a given space more strongly. For electro-magnetic inertia is wholly due to the magnetization of the space surrounding a current, and this space may be increased, or its magnetization intensified, as much as we please.

To increase the space we have only to make the discharge take a long circuit instead of a short one. Thus we may send it by a wire all round the room, or by a telegraph wire all round a town, and all the space inside it and some of that outside will be more or less magnetized. More or less, I say, and it is a case of less rather than more. Practically very little effect is felt except close to the conductor, and accordingly the self-induction increases very nearly proportionally to the length of the wire, and not in proportion to the area inclosed : provided also the going and return wires are kept a reasonable distance apart, so as not to encroach upon each other's appreciably magnetized regions. See Appendix (e).

But it is just as effective, and more compact, to intensify the magnetization of a given space by sending the current hundreds of times round it instead of only once ; and this is done by inserting a coil of wire into the discharge circuit.

Yet a third way there is of increasing the magnetization of a given space, and that is to fill it with some very magnetizable subtance such as iron. This, indeed, is a most powerful method under many circumstances, it being possible to increase the magnetization and therefore the self-induction or inertia of the current some 5000 times by the use of iron.

But in the case of the discharge of a Leyden jar iron is of no advantage. The current oscillates so quickly that any iron introduced into its circuit, however subdivided into thin wires it may be, is protected from magnetism by inverse currents induced in its outer skin, as your Professor of Natural

Philosophy [1] has shown, and accordingly it does not
get magnetized ; and so far from increasing the
inductance of the discharge circuit it positively
diminishes it by the reaction effect of these induced
currents : it acts, in fact, much as a mass of copper
might be expected to do.

The conditions determining rate of oscillation
being understood, we have next to consider what
regulates the damping out of the vibrations, *i.e.* the
total duration of the discharge.

Resistance is one thing. To check the oscillations
of a vibrating spring you apply to it friction, or
make it move in a viscous medium, and its vibrations
are speedily damped out. The friction may be made
so great that oscillations are entirely prevented, the
motion being a mere dead-beat return to the position
of equilibrium ; or, again, it may be greater still, and
the motion may correspond to a mere leak or slow
sliding back, taking hours or days for its accomplish-
ment. With very large condensers, such as are used
in telegraphy, this kind of discharge is frequent, but
in the case of a Leyden jar discharge it is entirely
exceptional. It can be caused by including in the
circuit a wet string, or a capillary tube full of distilled
water, or a slab of wood, or other atrociously bad
conductor of that sort ; but the conditions ordinarily
associated with the discharge of a Leyden jar,

[1] Lord Rayleigh.

whether it discharge through a long or a short wire, or simply through its tongs, or whether it overflow its edge or puncture its glass, are such as correspond to oscillations, and not to leak. [Discharge jar first through wire and next through wood.]

When the jar is made to leak through wood or water the discharge is found to be still not steady : it is not oscillatory indeed, but it is intermittent. It occurs in a series of little jerks, as when a thing is made to slide over a resined surface. The reason of this is that the terminals discharge faster than the circuit can supply the electricity, and so the flow is continually stopped and begun again.

Such a discharge as this, consisting really of a succession of small sparks, may readily appeal to the eye as a single flash, but it lacks the noise and violence of the ordinary discharge ; and any kind of moving mirror will easily analyze it into its constituents and show it to be intermittent. [Shake a mirror, or waggle head or opera-glass.]

It is pretty safe to say, then, that whenever a jar discharge is not oscillatory it is intermittent, and when not intermittent is oscillatory. There is an intermediate case when it is really dead-beat, but it could only be hit upon with special care, while its occurrence by accident must be rare.

So far I have only mentioned resistance or friction as the cause of the dying out of the vibrations ; but there is another cause, and that a most exciting one.

The vibrations of a reed are damped partly indeed by friction and imperfect elasticity, but partly also by the energy transferred to the surrounding medium and consumed in the production of sound. It is the formation and propagation of sound-waves which largely damp out the vibrations of any musical instrument. So it is also in electricity. The oscillatory discharge of a Leyden jar disturbs the medium surrounding it, carves it into waves which travel away from it into space : travel with a velocity of 185,000 miles a second : travel precisely with the velocity of light. [Tuning-fork.]

The second cause, then, which damps out the oscillations in a discharge circuit is *radiation :* *electrical* radiation if you like so to distinguish it, but it differs in no respect from ordinary radiation (or "radiant heat" as it has so often been called in this place) ; it differs in no respect from Light except in the physiological fact that the retinal mechanism, whatever it may be, responds only to waves of a particular, and that a very small, size, while radiation in general may have waves which range from 10,000 miles to a millionth of an inch in length.

The seeds of this great discovery of the nature of light were sown in this place : it is all the outcome of Faraday's magneto-electric and electrostatic induction : the development of them into a rich and full-blown theory was the greatest part of the life-work of Clerk Maxwell : the harvest of experimental verification is now being reaped by a German. But by no ordinary German. Dr. Hertz, now Professor in the University of Bonn, is a young investigator of the highest type. Trained in the school of Helmholtz, and endowed with both mathematical knowledge and great - experimental skill, he has immortalized himself by a brilliant series of investigations which have cut right into the ripe corn of scientific opinion in these islands, and by the same strokes as have harvested the grain have opened up wide and many branching avenues to other investigators.

At one time I had thought of addressing you this evening on the subject of these researches of Hertz, but the experiments are not yet reproducible on a scale suited to a large audience, and I have been so closely occupied with some not wholly dissimilar, but independently conducted, researches of my own— researches led up to through the unlikely avenue of lightning-conductors—that I have had as yet no time to do more than verify some of them for my own edification (§ 189).

In this work of repetition and verification Prof. Fitzgerald has, as related in a recent number of NATURE (vol. xxxix. p. 391), probably gone further ; and if I may venture a suggestion to your Honorary

Secretary, I feel sure that a discourse on Hertz's researches from Prof. Fitzgerald next year would be not only acceptable to you, but would be highly conducive to the progress of science.

I have wandered a little from my Leyden jar, and I must return to it and its oscillations. Let me very briefly run over the history of our knowledge of the oscillatory character of a Leyden jar discharge. It was first clearly realized and distinctly stated by that excellent experimentalist, Joseph Henry, of Washington, a man not wholly unlike Faraday in his mode of work, though doubtless possessing to a less degree that astonishing insight into intricate and obscure phenomena ; wanting also in Faraday's circumstantial advantages.

This great man arrived at a conviction that the Leyden jar discharge was oscillatory, by studying the singular phenomena attending the magnetization of steel needles by a Leyden jar discharge, first observed in 1824 by Savary. Fine needles, when taken out of the magnetizing helices, were found to be not always magnetized in the right direction, and the subject is referred to in German books as " anomalous magnetization." It is not the magnetization which is anomalous, but the currents which have no simple direction ; and we find in a memoir published by Henry in 1842, the following words :—

B B

" This anomaly, which has remained so long unexplained, and which, at first sight, appears at variance
with all our theoretical ideas of the connection of
electricity and magnetism, was, after considerable
study, satisfactorily referred by the author to an
action of the discharge of the Leyden jar which had
never before been recognized. The discharge, whatever may be its nature, is not correctly represented
(employing for simplicity the theory of Franklin) by
the single transfer of an imponderable fluid from one
side of the jar to the other ; the phenomenon requires
us to admit *the existence of a principal discharge in one
direction and then several reflex actions backward and
forward, each more feeble than the preceding, until the
equilibrium is obtained.* All the facts are shown to
be in accordance with this hypothesis, and a ready
explanation is afforded by it of a number of phenomena, which are to be found in the older works
on electricity, but which have until this time
remained unexplained." [1]

The italics are Henry's. Now if this were an
isolated passage it might be nothing more than a
lucky guess. But it is not. The conclusion is one
at which he arrives after a laborious repetition and
serious study of the facts, and he keeps the idea constantly before him when once grasped, and uses it in

[1] *Scientific Writings of Joseph Henry,* vol. i. p. 201. Published
by the Smithsonian Institution, Washington, 1886.

all the rest of his researches on the subject. The facts studied by Henry do in my opinion support his conclusion, and if I am right in this it follows that he is the original discoverer of the oscillatory character of a spark, although he does not attempt to state its theory. That was first done, and completely done, in 1853, by Sir William Thomson ; and the progress of experiment by Feddersen, Helmholtz, Schiller, and others has done nothing but substantiate it.

The writings of Henry have been only quite recently collected and published by the Smithsonian Institution of Washington in accessible form, and accordingly they have been far too much ignored. The two volumes contain a wealth of beautiful experiments clearly recorded, and well repay perusal.

The discovery of the oscillatory character of a Leyden jar discharge may seem a small matter, but it is not. One has only to recall the fact that the oscillators of Hertz are essentially Leyden jars—one has only to use the phrase " electro-magnetic theory of light "—to have some of the momentous issues of this discovery flash before one.

One more extract I must make from that same memoir by Henry,[1] and it is a most interesting one : it shows how near he was, or might have been, to obtaining some of the results of Hertz ; though, if he had obtained them, neither he nor any other experi-

[1] *Loc. cit.*, p. 204.

mentalist could possibly have divined their real significance.

It is, after all, the genius of Maxwell and of a few other great theoretical physicists whose names are on everyone's lips [1] which endows the simple induction experiments of Hertz and others with such stupendous importance.

Here is the quotation :—

" In extending the researches relative to this part of the investigations, a remarkable result was obtained in regard to the distance at which induction effects are produced by a very small quantity of electricity ; a single spark from the prime conductor of a machine, of about an inch long, thrown on to the end of a circuit of wire in an upper room, produced an induction sufficiently powerful to magnetize needles in a parallel circuit of iron placed in the cellar beneath, at a perpendicular distance of 30 feet, with two floors and ceilings, each 14 inches thick, intervening. The author is disposed to adopt the hypothesis of an electrical *plenum* " [in other words, of an ether], " and from the foregoing experiment it would appear that a single spark is sufficient to disturb perceptibly the electricity

---

[1] And of one whose name is not yet on everybody's lips, but whose profound researches into electro-magnetic waves have penetrated further than anybody yet understands into the depths of the subject, and whose papers have very likely contributed partly to the theoretical inspiration of Hertz—I mean that powerful mathematical physicist, Mr. Oliver Heaviside.

of space throughout at least a cube of 400,000 feet of
capacity; and when it is considered that the magnetism
of the needle is the result of the difference of two
actions, it may be further inferred that the diffusion of
motion in this case is almost comparable with that of
a spark from a flint and steel in the case of light."

Comparable it is, indeed, for we now know it to be
the self-same process.

One immediate consequence and easy proof of the
oscillatory character of a Leyden jar discharge is the
occurrence of phenomena of sympathetic resonance.

Everyone knows that one tuning-fork can excite
another at a reasonable distance if both are tuned to
the same note.   Everyone knows, also, that a fork can
throw a stretched string attached to it into sympathetic
vibration if the two are tuned to unison or to some
simple harmonic.   Both these facts have their electrical
analogue.   I have not time to go fully into the matter
to night, but I may just mention the two cases which
I have myself specially noticed.

A Leyden jar discharge can so excite a similarly-
timed neighbouring Leyden jar circuit as to cause the
latter to burst its dielectric if thin and weak enough.
The well-timed impulses accumulate in the neighbouring circuit till they break through a quite perceptible
thickness of air.

Put the circuits out of unison, by varying the capacity
or by including a longer wire in one of them ; then,

although the added wire be a coil of several turns, well adapted to assist mutual induction as ordinarily under-stood, the effect will no longer occur. It can be obtained again by diminishing the static capacity.

That is one case, and it is the electrical analogue of one tuning-fork exciting another. It is too small at present to show here satisfactorily, for I only recently observed it, but it is exhibited in the library at the back.

The other case, analogous to the excitation of a stretched string of proper length by a tuning-fork, I published last year under the name of the experiment of the recoil kick; where a Leyden jar circuit sends waves along a wire connected by one end with it, which waves splash off at the far end with an electric brush or long spark.

I will show merely one phase of it to-night, and that is the reaction of the impulse accumulated in the wire upon the jar itself, causing it to either overflow or burst. [Sparks of gallon or pint jar made to over-flow by wire round room.[1]]

---

[1] During the course of this experiment, the gilt paper on the wall was observed by the audience to be sparkling, every gilt patch over a certain area discharging into the next, after the manner of a spangled jar. It was probably due to some kind of sympathetic resonance. Electricity splashes about in conductors in a surprising way everywhere in the neighbourhood of a discharge. For instance, a telescope in the hand of one of the audience was reported afterwards to be giving off little sparks at every discharge of the jar. Everything which happens to have a period of electric oscillation corresponding to some harmonic of the main oscillation of a discharge is liable to behave in this way. When light falls on an opaque surface it is quenched; producing minute electric currents, which subside into heat. What the audience saw was

The early observations by Franklin on the bursting of Leyden jars, and the extraordinary complexity or multiplicity of the fracture that often results, are most interesting. (See *Electrician* for March 29 and April 5, 1889.)

His electric experiments as well as Henry's well repay perusal, though of course they belong to the infancy of the subject.

He notes the striking fact that the bursting of a jar is an extra occurrence—it does not replace the ordinary discharge in the proper place, it accompanies it ; and we now know that it is precipitated by it, that the spark occurring properly between the knobs sets up such violent surgings that the jar is far more violently strained than by the static charge or mere difference of potentials between its coatings ; and if the surgings are at all even roughly properly timed, the jar is bound to either overflow or burst.

Hence a jar should always be made without a lid, and with a lip protruding a carefully considered distance above its coatings : not so far as to fail to act as a safety valve, but far enough to prevent overflow under ordinary and easy circumstances.

probably the result of waves of electrical radiation being quenched or reflected by the walls of the room, and generating electrical currents in the act (§ 166). It is these electric surgings which render such severe caution necessary in the erection of lightning-conductors.

This explanation has since been entirely confirmed by similar occurrences in other places.

And now we come to what is after all the main
subject of my discourse this evening, viz. the optical
and audible demonstration of the oscillations occurring in the Leyden jar spark. Such a demonstration
has, so far as I know, never before been attempted, but
if nothing goes wrong we shall easily accomplish it.

And first I will do it audibly. To this end the
oscillations must be brought down from their extraordinary frequency of a million or hundred thousand
a second to a rate within the limits of human audition.
One does it exactly as in the case of the spring—one
first increases the flexibility and then one loads it.
[Spark from battery of jars and varying sound of
same.]

Using the largest battery of jars at our disposal, I
take the spark between these two knobs—not a long
spark, ¼ inch will be quite sufficient. Notwithstanding
the great capacity, the rate of vibration is still far
above the limit of audibility, and nothing but the
customary crack is heard. I next add inertia to the
circuit by including a great coil of wire, and at once
the spark changes character, becoming a very shrill
but an unmistakable whistle, of a quality approxi-
mating to the cry of a bat. Add another coil, and
down comes the pace once more, to something like
5000 per second, or about the highest note of a piano.
Again and again I load the circuit with magnetiza-
bility, and at last the spark has only 500 vibrations

a second, giving the octave, or perhaps the double octave, above the middle C.

One sees clearly why one gets a musical note : the noise of the spark is due to a sudden heating of the air ; now if the heat is oscillatory, the sound will be oscillatory too, but both will be an octave above the electric oscillation, if I may so express it, because two heat-pulses will accompany every complete electric vibration, the heat production being independent of direction of current.

Having thus got the frequency of oscillation down to so manageable a value, the optical analysis of it presents no difficulty : a simple looking-glass waggled in the hand will suffice to spread out the spark into a serrated band, just as can be done with a singing or a sensitive flame : a band too of very much the same appearance.

Using an ordinary four-square rotating mirror driven electro-magnetically at the rate of some two or three revolutions per second, the band is at the lowest pitch seen to be quite coarsely serrated ; and fine serrations can be seen, with four revolutions per second, in even the shrill whistling sparks.

The only difficulty in seeing these effects is to catch them at the right moment. They are only visible for a minute fraction of a revolution, though the band may appear drawn out to some length. The further away a spark is from the mirror, the more drawn out it is, but also the less chance there is of catching it.

With a single observer it is easy to arrange a contact maker on the axle of the mirror which shall bring on the discharge at the right place in the revolution, and the observer may then conveniently watch for the image in a telescope or opera-glass; though at the lower pitches nothing of the kind is necessary.

But to show it to a large audience various plans can be adopted   One is to arrange for several sparks instead of one; another is to multiply images of a single spark by suitably adjusted reflectors, which if they are concave will give magnified images; another is to use several rotating mirrors; and indeed I do use two, one adjusted so as to suit the spectators in the gallery.

But the best plan that has struck me is to combine an intermittent and an oscillatory discharge. Have the circuit in two branches, one of high resistance so as to give intermittences, the other of ordinary resistance so as to be oscillatory, and let the mirror analyze every constituent of the intermittent discharge into a serrated band. There will thus be not one spark, but a multitude of successive sparks, close enough together to sound almost like one, separate enough in the rotating mirror to be visible on all sides at once.

But to achieve it one must have great exciting power. In spite of the power of this magnificent

Wimshurst machine, it takes some time to charge up our great Leyden battery, and it is tedious waiting for each spark. A Wimshurst does admirably for a single observer, but for a multitude one wants an instrument which shall charge the battery not once only but many times over, with overflows between, and all in the twinkling of an eye.

To get this I must abandon my friend Mr. Wimshurst, and return to Michael Faraday. In front of the table is a great induction coil ; its secondary has the resistance needed to give an intermittent discharge. The quantity it supplies at a single spark will fill our jars to overflowing several times over. The discharge circuit and all its circumstances shall remain unchanged. [Excite jars by coil.]

Running over the gamut with this coil now used as our exciter instead of the Wimshurst machine—everything else remaining exactly as it was—you hear the sparks give the same notes as before, but with a slight rattle in addition, indicating intermittence as well as alternation. Rotate the mirror, and everyone should see one or other of the serrated bands of light at nearly every break of the primary current of the coil. [Rotating mirror to analyze sparks.]

The musical sparks which I have now shown you

were obtained by me during a special digression [1] which I made while examining the effect of discharging a Leyden jar round heavy glass or bisulphide of carbon. The rotation of the plane of polarization of light by a steady current, or by a magnetic field of any kind properly disposed with respect to the rays of light, is a very familiar one in this place. Perhaps it is known also that it can be done by a Leyden jar current. But I do not think it is ; and the fact seems to me very interesting. It is not exactly new—in fact, as things go now it may be almost called old, for it was investigated six or seven years ago by two most highly skilled French experimenters, Messrs. Bichat and Blondlot.

But it is exceedingly interesting as showing how short a time, how absolutely no time, is needed by heavy glass to throw itself into the suitable rotatory condition. Some observers have thought they had proved that heavy glass requires time to develop the effect, by spinning it between the poles of a magnet and seeing the effect decrease ; but their conclusions cannot be right, for the polarized light follows every oscillation in a discharge, the plane of polarization being waved to and fro as often as 70,000 times a second in my own observation. (See *Phil. Mag.* April 1889.)

[1] Most likely it was a conversation which I had with Sir Wm. Thomson, at Christmas, which caused me to see the interest of getting slow oscillations. My attention has mainly been directed to getting them quick.

Very few persons in the world have seen the effect. In fact, I doubt if anyone had seen it a month ago except Messrs. Bichat and Blondlot. But I hope to make it visible to most persons here, though I hardly hope to make it visible to all.

Returning to the Wimshurst machine as exciter, I pass a discharge round the spiral of wire inclosing this long tube of CS_2, and the analyzing Nicol being turned to darkness, there may be seen a faint—by those close to not so faint, but a very momentary— restoration of light on the screen at every spark. [CS_2 tube experiment on screen.]

Now I say that this light restoration is also oscillatory. One way of proving this fact is to insert a biquartz between the Nicols. With a steady current it constitutes a sensitive detector of rotation, its sensitive tint turning green on one side and red on the other. But with this oscillatory current a biquartz does absolutely nothing. [Biquartz.]

That is one proof. Another is that rotating the analyzer either way weakens the extra brightening of the field, and weakens it equally either way.

But the most convincing proof is to reflect the light coming through the tube upon our rotating mirror, and to look now not at the spark, or not only at the spark, but at the faint band into which the last residue of light coming through polarizer and tube and analyzer is drawn out. [Analyze the light in rotating mirror.]

At every discharge this faint streak brightens in places into a beaded band : these are the oscillations of the polarized light ; and when examined side by side they are as absolutely synchronous with the oscillations of the spark itself as can be perceived.

Out of a multitude of phenomena connected with the Leyden jar discharge I have selected a few only to present to you here this evening. Many more might have been shown, and great numbers more are not at present adapted for presentation to an audience, being only visible with difficulty and close to.

An old and trite subject is seen to have in the light of theory an unexpected charm and brilliancy. So it is with a great number of other old familiar facts at the present time.

The present is an epoch of astounding activity in physical science. Progress is a thing of months and weeks, almost of days. The long line of isolated ripples of past discovery seem blending into a mighty wave, on the crest of which one begins to discern some oncoming magnificent generalization. The suspense is becoming feverish, at times almost painful. One feels like a boy who has been long strumming on the silent keyboard of a deserted organ, into the chest of which an unseen power begins to blow a vivifying breath. Astonished, he now finds that the touch of a finger elicits a responsive note, and he hesitates, half delighted, half affrighted, lest he be deafened by the chords which it would seem he can now summon forth almost at will.

APPENDIX.

APPENDIX.

CERTAIN portions of electrical science have recently come into considerable prominence, and, as they are hardly satisfactorily treated in text-books yet, it may be a help to students to say something about them here in less popular language than in the body of the book.

Electro-magnetism.

(*a*) The fundamental fact of electro-magnetism, ascertained by direct experiment, is that a circuit conveying a current exactly imitates a magnet of definite moment, the equivalent moment being

$$ml = \mu n \mathrm{A C},$$

where A is the mean area of the coil, n the number of turns of wire, C the current, and μ a constant characteristic of the medium inside the coil, whose absolute value we have as yet no means of ascertaining (§§ 68, 69, 127).

Magnetic Induction, Reluctance, and Permeability.

(*b*) The intensity of magnetic field at a distance r from a pole of strength m is $\frac{m}{r^2}$, and this may be called the number of lines of force (or tubes if the idea be preferred) per unit area. The total number of lines of force through a spherical surface of this radius is $\frac{m}{r^2} \times 4 \pi r^2$, or $4 \pi m$.

This number must likewise thread any closed surface whatever inclosing the pole ; and in fact it is the number the pole
possesses.   It may be called the total magnetic flux or displacement, or the total magnetic induction, due to the pole ; the
name " induction," first used vaguely in the sense of *influence* by
Faraday, having been given this definite connotation by Maxwell.
The same expression likewise gives the number of lines of force
due to a complete magnet ; for the superposition of lines due to an
equal opposite pole curves the original lines but alters not their
number.   With two detached poles the lines simply go from
one to the other.   With a complete magnet the lines all form
closed loops extending from north to south through air, and
back through steel.   In the case of a coil they likewise are
closed loops, all threading the coil and then spreading out
through the surrounding medium.   In all real cases, therefore,
the lines of force form closed curves.   Magnetic circuits are
always closed, just as electric circuits are.

Take the simplest case of an anchor-ring coil, a helix bent
into a closed circuit (like Fig. 47 or 29) : all its lines are then
inside it, and their total number, being $4\pi m$, is $\dfrac{4\pi\mu nAC}{l}$; where

l is the mean circumference of the anchor-ring, or length of the
magnetic circuit.   This is called the total flux of magnetic
induction, or briefly the total induction, and we will denote it
by I.

Now, in the analogous case of a voltaic circuit, the current
is ratio of electromotive force to resistance, and the resistance
may be written $\dfrac{l}{\kappa A}$; κ being specific conductivity, and A sectional
area of conductor of length l.

To bring out the analogy, we shall write the magnetic flux—

$$I = \frac{4\pi nC}{\dfrac{l}{\mu A}},$$

where the numerator is sometimes called magneto-motive force, and the denominator magnetic resistance, or preferably, as suggested by Mr. Heaviside, magnetic *reluctance*. Obviously μ takes the place of electric conductivity, and is a sort of magnetic conductivity : it was from this point of view that Sir W. Thomson long ago christened it "permeability" (see § 82).

If the magnetic circuit is not so simply constituted, but is composed of portions of different areas, length, and material in series—as the magnetic circuit of a dynamo is, for instance—the magnetic reluctance can be written (still pursuing the analogy)—

$$R = \frac{l_1}{\mu_1 A_1} + \frac{l_2}{\mu_2 A_2} + \ldots,$$

and $I = \dfrac{4\pi n C}{R}$ as before.

Mutual Induction.

(*c*) If a single turn of secondary wire surround this closed magnetic circuit, as in Fig. 47, the total induction through it, whatever its shape or size, is just I ; and if it surround the ring n' times, the effective total induction is n'I. This is the induction of the primary through the secondary, which, written out in full, is—

$$\frac{4\pi \mu n n' A C}{l}.$$

The relation is a mutual one ; and if the same current were to flow in secondary, the same number of lines would thread effectively the primary. Hence we call it *mutual* induction, and write it MC ; where M, the coefficient of mutual induction between the two coils, is—

$$M = \frac{4\pi \mu n n' A}{l} ;$$

the A and the l referring most easily to the simply and obviously closed magnetic circuit. Two detached coils situated anyhow with respect to each other, will have a specifiable value of M, but it is not so easy to write down.

Self-Induction.

(d) Instead of using a secondary coil to surround the induction caused by the primary, we may consider the primary as surrounding the induction itself has produced, and so speak of its "self-induction" as—

$$\frac{4\pi\mu n^2 A C}{l},$$

which, written LC, gives us the coefficient of self-induction—

$$L = \frac{4\pi\mu n^2 A}{l},$$

or,

$$= 4\pi\mu n n_1 A,$$

where n_1 = number of turns per unit length. (§§ 115 and 98)

Here, again, every coil has a specifiable self-induction, but in most cases it is not so easy to write down. It always means, however, the ratio of the self-produced magnetic induction to the current which has produced it—

$$L = \frac{I}{C}.$$

Value of Coefficient of Self-Induction in a few other Simple Cases.

(e) The magnetic field produced by a straight wire varies inversely with the distance; being, at a distance r from a straight wire of sectional radius a, conveying a current, C—

$$\frac{2\mu C}{r}.$$

and this therefore specifies the number of lines through unit area.

So the whole number of lines of force included between the wire and any distance b, in a drum of thickness l, is—

$$\int_a^b \frac{2\mu C l}{r}\, dr = 2\mu C l \cdot \log \frac{b}{a}.$$

Now, if at the distance b there is a parallel wire, conveying the return current, it, too, will have the same number of lines of force, and the whole number lying between a length, l, of each of the two parallel wires is—

$$4\mu l \log \frac{b}{a} \times C ;$$

and as all the lines of force that exist pass between the wires, this expression sums up the whole magnetic flux produced by the going and return parallel currents; and the co-efficient of C in the last expression is therefore the coefficient of self-induction for the case of two thin parallel wires at a distance b.

For a circular loop of radius r, radius of section of wire being a, this modifies itself to—

$$L = 4\pi\mu r \log \frac{8r}{a}$$

(see § 140). In every case μ refers to the space near the wire, not to the substance of the wire itself.

In both these cases, the magnetization of the substance of the wires themselves is supposed *nil*. In the case of extremely rapidly alternating currents, this is correct (§ 47). In the case of copper wires not too close together, it is never *very* incorrect.

Energy of a Current.

(f) A magnet of moment ml, in a magnetic field of intensity H, experiences a couple mlH $\sin \theta$; and therefore a simple stiff coil of wire conveying a current experiences a couple μnACH sin θ. If it turns a small angle, $d\theta$, the work done, or the change of potential energy, is μnACH sin $\theta\, d\theta$; and therefore the potential energy of the circuit in any position is $-\mu n$ACH cos θ ; which may be written IC, because nA cos θ is the effective area of the coil resolved perpendicularly to the lines of force which thread it to the number μH per unit area.

This result may be generalized ; a current in a magnetic field always possesses energy IC. If the field is due to external causes, *i.e.* having an existence independent of the current, the energy is potential energy of strain, and tends to cause the circuit to rotate. This is the principle of electric motors. But if the field is due to nothing but the current itself—if it is a self-produced and self-maintained field—the value of I is LC, and the energy is now more conveniently called kinetic energy. To obtain its value, we must remember that the induction and the current die out together : it is not as if they had an independent existence, and so the energy is—

$$\int_0^c I dC = \tfrac{1}{2}LC^2.$$

This is the work which must be done at starting and at stopping the current (Chap. V.).

Pole near a Circuit.

(*g*) If a single pole find itself on the axis of a circle, the number of its lines of force which penetrate the circle is $\frac{m}{r^2} \cdot 2\pi r^2(1 - \cos\theta)$, the latter factor being the area of the portion of a sphere with centre at m, cut off by the said circle. The expression $2\pi(1 - \cos\theta)$, since it measures the ratio of the area subtended by a conical angle to the square of the radius, is, in analogy with the circular measure of a plane angle, called a solid angle : the solid angle of the cone with vertex m and base the circle, or the angle subtended by the circle to an eye placed at m. Call this angle ω; then the number of lines of force, or the magnetic induction through the circle is $m\omega$.

If the circle becomes now a circuit conveying a current C, the system has energy $m\omega$C, and accordingly there will be a tendency to relative motion, the force in any direction being equal to the rate of change of $m\omega$C per unit distance in that direction.

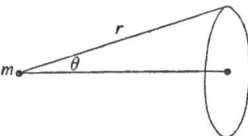

The potential of the pole on the circuit is $m\omega$; the potential of the circuit on the pole is Cω. If the pole is situated anywhere, and the coil is of any shape, ω can still be specified, but not so easily. If there is a collection of magnets, their potential on a circuit, or induction through it, can be written $\Sigma(m\omega)$.

Magneto-electricity.

(h) The fundamental fact of magneto-electricity is that if the induction through a circuit change from any cause whatever, an E.M.F. is set up in the circuit equal to the rate of change of the magnetic induction—

$$e = \frac{d\mathrm{I}}{dt}.$$

This is not strictly a relation independent of the fundamental fact of electro-magnetism : the two are connected by the law of the conservation of energy. I may indicate this important

394   APPENDIX.

fact sufficiently for our present purpose by quoting the conservation of energy, in a form applicable to the case of a circuit conveying a steady current, as—

$$E C dt = R C^2 dt + C d I ;$$

whence

$$R C = E - \frac{d I}{d t},$$

or the resultant E.M.F. consists not only of the E.M.F. applied, but contains also an intrinsic or indirect E.M.F. magnetically excited in the circuit ; this being what Faraday discovered as magneto-electricity.

Various Modes of exciting Induction Currents.

(i) Now I may be made up in a multitude of ways. It may be a component of terrestrial magnetic field, say, $n A H \cos \theta$. It may be caused by magnets in the neighbourhood $\Sigma(m\omega)$. It may be due to induction from some other coil, MC. It may be due to the current passing in the coil itself, say LC. The total induced E.M.F. is the rate of change of the sum of all these, or—

$$e = \frac{d}{d t} \{ n A H \cos \theta + \Sigma(m\omega) + M C' + L C \} ;$$

and accordingly it may be excited in many ways: by changes in size or shape of coil; by changing its aspect to the field (as in a dynamo); by moving magnets in its neighbourhood (as in an alternating-current machine); by varying the current in or shifting the position of other circuits (as in a Ruhmkorff coil); or, lastly, by changing its own current, or its own coefficient of self-induction. Changes in the last term, $\frac{d}{d t} (L C)$, are specially called E.M.F. of self-induction, and used to be called extra-currents.

Primary Current alone : and Coil with Revolving Commutator.

(*j*) The equation to a current of varying strength in the simplest case of a lone circuit is—

$$E - RC = \frac{d}{dt}(LC),$$

where E is the applied E.M.F.; and this may be written out more fully—

$$L\frac{dC}{dt} + \left(R + \frac{dL}{dt}\right)C = E,$$

which shows that in the case of circuits of variable self-induction the resistance has not its most simple value, but has an extra term in it, a spurious or imitative resistance, $\frac{dL}{dt}$.

An example of a circuit of variable self-induction is one which is continually having wire withdrawn from or added to it, so that a current has to be stopped in portions where it was already established, and started in hitherto stagnant portions : a case quite analogous to the viscosity of gases, and commonly illustrated by passengers of appreciable inertia getting in and out of a moving train. An instance of the case occurs in every Gramme ring, or indeed every dynamo armature, when spinning with a commutator, quite independently of the magnetic field in which it may happen to be spinning. In all such cases the effective resistance is rather greater than R, being $R + \frac{dL}{dt}$, or $R + nL$; where the self-induction virtually added to the circuit n times a second is L.

Leyden Jar.

(*k*) In the case of a discharging condenser of capacity S, the quantity stored in it at any instant is such that $C = -\frac{dQ}{dt}$, or that $Q = Q_0 - \int_0^t C dt$; and the difference of potential be-

tween its terminals is $\frac{Q}{S}$, which is the E.M.F. applied to the

circuit. So the equation to the discharge current is —

$$L\frac{dC}{dt} + RC = \frac{Q}{S}.$$

The solution of the equation in this case is—

$$C = \frac{E}{pL} e^{-mt} \sin pt,$$

where $m = \frac{-R}{2L}$, and regulates the total duration of the discharge,

and where $p = \frac{1}{\sqrt{(LS)}}$ approximately

$$\left\{ \text{more accurately } \sqrt{\left(\frac{1}{LS} - m^2\right)} \right\},$$

and regulates the rapidity of alternation, which is $\frac{p}{2\pi}$. The
wave-length of the emitted radiation (Chapter XIV.) is—

$$\lambda = \frac{2\pi}{p} . v = 2\pi \sqrt{\left(\frac{L}{\mu} . \frac{S}{K}\right)}.$$

With these quick oscillations, R is nothing at all like its
ordinary value for steady currents ; because the outside of the
wire only is used (§§ 45 and 102) ; but, calling the ordinary value
R_0, R is very approximately, for high rates of alternation,—[1]

$$R = \sqrt{(\tfrac{1}{2}p\mu_0 l \cdot R_0)},$$

l being the length of the wire, and μ_0 the magnetic permeability
of its *substance* (§ 46).

---

[1] See Rayleigh, *Phil. Mag.*, May 1886.

The emission of radiation by such a circuit goes to increase R still more (§ 142 and p. 367). See also *m*.

Alternating Current.

(*l*) In case of any coil or armature spinning in a magnetic field, the equation to the current is—

$$- \text{RC} = \frac{d}{dt}\Big(n\text{AH}\cos\theta + \text{LC}\Big),$$

$$\text{or } \text{L}\frac{d\text{C}}{dt} + \Big(\text{R} + \frac{d\text{L}}{dt}\Big)\text{C} = n\text{AH}\sin\theta\,\frac{d\theta}{dt};$$

and the E.M.F. is therefore alternating according to a sine function. Writing this equation—

$$\text{L}\frac{d\text{C}}{dt} + \text{R}'\text{C} = \text{E}_0\sin pt,$$

the solution is—

$$\text{C} = \frac{\text{E}_0\cos(pt - \epsilon)}{\sqrt{\{\text{R}^2 + (p\text{L})^2\}}},$$

where $\tan\epsilon = \frac{p\text{L}}{\text{R}'}$. The R' differs from simple R, as already explained in (*j*), only when a commutator is employed : which it often is not. The denominator of the above expression

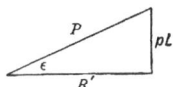

may be called impēdance, and denoted by P (see next section), the quantities being related as in this little diagram. The quantity ϵ is the lag of the current behind the applied E.M.F.

Two Definitions of Electric Resistance, and Distinction between the Two.

(*m*) The oldest definition of the term " resistance of a conductor " is that given by Ohm, viz. the ratio—

$$\frac{\text{E.M.F. applied to the conductor}}{\text{Current excited in it}}.$$

But another is contained in the law of Joule, viz. the ratio—

$$\frac{\text{Energy dissipated per second by the conductor}}{\text{Current squared which it transmits}}.$$

In cases of no reversible obstruction the two definitions agree, but in cases of chemical action, of reversible heat effects, and of varying magnetic induction, some of the energy may be stored, all is not dissipated, and under these circumstances the two definitions do not agree. A distinction must be drawn between them : the term resistance cannot properly be applied to both quantities.

Now it is found convenient to retain the name resistance for the second definition—the dissipation of energy coefficient ; and to realize that in the total obstruction specified by the first definition there is included " back E.M.F.," " polarization," or other reversible obstruction, in addition to resistance proper ; while in the very important case of the total obstruction met with by an alternating current, it has become convenient to call the quantity defined by the first of the two equations, " impēdance."

The two definitions of resistance may indeed be always made to agree, if, in the Ohm's law definition, instead of *applied* E.M.F., we reckon *resultant* E.M.F. And this is the neatest and simplest mode of taking into account such things as chemical or thermal polarization, and also a magnetic back E.M.F., so long as it is steady and external, as in the case of electric motors. But, when dealing with alternating generators, some understanding has to be come to as to how the value of their

E.M.F. is to be reckoned, and no simple subtraction of a back E.M.F. is convenient. Referring to last section, we see that the expression for current contains as numerator a lessened or lagging E.M.F., and as denominator an obstruction or impēdance containing a term in addition to what is usually called resistance. It is from this point of view that the idea and term "impēdance" become so useful.

The value of this quantity is, in general, as has been shown,

$$\sqrt{\{(pL)^2 + R^2\}} \; ;$$

and its two portions may be styled respectively the inertia, or conservative portion, and the frictional or dissipative portion (§ 38).

Part of the energy dissipated appears as heat in the conductor, and this is the only portion on which Joule experimented, but another portion we now know is propagated out as radiation into space (§ 142) : both portions together are included in the numerator proper to the second definition of R.

Induced Current in Secondary Circuit. Transformers.

(*n*) The E.M.F. induced in a secondary circuit surrounding a ring like Fig. 47, whose primary coil has an alternating or intermittent current, C, sent round it, is, referring back to (*h*) and (*c*)—

$$M \frac{dC}{dt}, \quad \text{or} \quad 4\pi n n' \; \frac{\mu A}{l} \cdot \frac{dC}{dt} \; ;$$

and depends, therefore, directly on the number of turns of wire in the secondary coil, and on the rate of variation of the primary current. This is the principle of induction-coils, and of "secondary-generators" or transformers (§ 115). The E.M.F. thus obtained is completely under control by choosing a suitable value for *n'*, according as high E.M.F. (in Ruhmkorff coils) or a powerful current (for electric welding) is required. They are called transformers, because, of the two electrical factors in mechanical "power," EC, they can change their ratio, leaving the product nearly constant ; just as ordinary machines do with the force and velocity factors of the same product "power." So, in precise analogy with gaining in force what you lose in speed, you gain in E.M.F. what you lose in current ; or *vice versâ*.

The equations to primary and secondary currents, C and C , are—

$$E - RC = \frac{d}{dt}\left(LC + MC'\right),$$

$$o - R'C' = \frac{d}{dt}\left(L'C' + MC\right);$$

and from the solution of these, the effective or apparent self-induction of primary, when its secondary is short-circuited and when all resistances are kept small, comes out equal to $L - \frac{M^2}{L'}$. Now since, for a simply closed magnetic circuit,

$$L : L' : M = n^2 : n'^2 : nn',$$

the effective self-induction (and therefore the impēdance) of the primary is approximately zero when its secondary is short-circuited—a fact which is the Magna Charta of commercial transformers.

Rate of Transmission of Telegraph Signals, in the Simplest Case.

(*o*) Consider a unit length of a pair of parallel thin copper wires not very close together, a going and return wire, at a distance *b* apart, the sectional radius of each wire being *a*. The self-induction of this portion, see (*e*), is—

$$L_1 = 4\mu \log \frac{b}{a},$$

and the static capacity of the same portion is (by somewhat similar reasoning)—

$$S_1 = \frac{K}{4 \log \frac{b}{a}}.$$

Hence

$$L_1 S_1 = \mu K.$$

The resistance of the same unit length may be called R_1.

Now consider an element of the pair of wires of length dx, and write down the slope of potential between its ends when a current, C, flows along it, and also their rise of potential with time ; we get—

$$L_1 \frac{dC}{dt} + R_1 C + \frac{dV}{dx} = 0,$$

and

$$S_1 \frac{dV}{dt} + \frac{dC}{dx} = 0.$$

Now, a "wave" being any disturbance periodic both in space and time, its general fundamental equation is—

$$y = a \sin (pt - nx),$$

where y is the extent of the disturbance at any place distant x from the origin, and at any time, t, from the era of reckoning.

The coefficient a is the amplitude of the vibration ; n is the space-period-constant, or $\frac{2\pi}{\lambda}$; p is the time-period-constant, or $\frac{2\pi}{T}$; the velocity of advance of the waves is one space-period in one time period, viz. $\frac{\lambda}{T}$ or $\frac{p}{n}$.

The solution of these equations for the case of an applied rapidly alternating E.M.F., V sin pt, at the origin, may be written—

$$V = V_0 e^{-\frac{m_1}{p_1} x} \sin p \left(t - \frac{x}{p_1} \right),$$

where $m_1 = \frac{R_1}{2L_1}$ and $p_1 = \frac{1}{\sqrt{(L_1 S_1)}}.$

Hence the above bracketed pair of equations give waves travelling along the wires with the speed $\dfrac{1}{\sqrt{(L_1 S_1)}}$, which we have seen equals $\dfrac{1}{\sqrt{(\mu K)}}$, and with an amplitude dying out along the length of the wires according to a logarithmic decrement $\tfrac{1}{2} R_1 \sqrt{\left(\dfrac{S_1}{L_1}\right)}$.

The speed of propagation of pulses along wires is therefore precisely the same, in this simple case, as the propagation of waves out through free space, viz. the velocity $\dfrac{1}{\sqrt{(\mu K)}}$ (§§ 128, 132, 137). All complications go to decrease, not to increase, the speed (§ 135).

Dimensions of Electrical Quantities.

(p) Writing L, M, T, F, v, for units of length, mass, time, force, velocity, as usual, and A for area ; the fundamental and certain experimental relations, independent of all considerations about units and systems of measurement, are—

Of electrostatics, $\quad Q = L \sqrt{(KF)}$ . . . . . . (1)
Of magnetism, $\quad\quad\; m = L \sqrt{(\mu F)}$ . . . . . (2)
Of electro-magnetism, $mL = \mu AC$ . . . . . . (3)
The last may also be written—
$$m = \mu v Q \quad\quad . . . . . . (3')$$
in which form it suggests the magnetic action of a moving charge, which Rowland's experiment has established.

Combining the three equations, we deduce—
$$\sqrt{\left(\frac{\mu}{K}\right)} = \frac{m}{Q} = \mu v ;$$
whence $\quad\quad \mu K = \dfrac{1}{v^2} = \dfrac{\text{density}}{\text{elasticity}},$

the well-known relation connecting the two etherial constants.

Comparing many electrical equations with corresponding mechanical ones, we find that the product LC takes the place of momentum (mv), and that $\frac{1}{2}LC^2$ takes the place of kinetic energy $(\frac{1}{2}mv^2)$, and indeed *is* the energy of a current, see (f). Hence it is natural to think of L as involving inertia, and of μ or $4\pi\mu$ as a kind of density of the medium concerned.

Assuming this, $\frac{4\pi}{K}$ at once becomes an elasticity coefficient (as indeed electrostatics itself suggests), because $\mu Kv^2 \equiv 1$; and the dimensions of all electrical units can be specified as follows, without any arbitrary convention or distinction between electrostatic and electro-magnetic units :—

Sp. ind. cap., $K = \dfrac{\text{strain}}{\text{stress}} = \dfrac{\text{area}}{\text{force}} = \dfrac{LT^2}{M} = $ shearability.

Permeability, $\mu = \dfrac{\text{inertia}}{\text{volume}} = \dfrac{M}{L^3} = $ density.

Electric charge, $Q = L^2 = \dfrac{\text{volume}}{\text{displacement}}$.

Magnetic pole, $m = \dfrac{M}{T} = $ momentum per unit length.

Electric current, $C = \dfrac{L^2}{T} = $ displacement $\times$ velocity.

Magnetic moment, $ml = \dfrac{ML}{T} = $ momentum.

E.M.F., $E = \dfrac{\text{work}}{Q} = \dfrac{M}{T^2} = $ pressure $\times$ displacement, or work per unit area.

Intensity of magnetic field, $H = \dfrac{F}{m} = \dfrac{L}{T} = $ velocity.

Intensity of electrostatic field, $\dfrac{F}{Q} = \dfrac{M}{LT^2} = $ energy per unit volume.

Surface density, $\sigma = \dfrac{Q}{A} = $ a pure number.

Electric tension, $\dfrac{2\pi\sigma^2}{K} = \dfrac{M}{LT^2} = $ a pressure or tension.

Capacity, $S = \dfrac{Q}{E} = \dfrac{L^2T^2}{M} =$ displacement per unit pressure.

Coefficient of resistance, $\dfrac{E}{C} = \dfrac{M}{L^2T} =$ impulse or momentum

per unit volume.

Magneto-motive force, $4\pi nC = \dfrac{L^2}{T} =$ current.

Reluctance, $\dfrac{l}{\mu A} = \dfrac{L^2}{M} = \dfrac{\text{area}}{\text{inertia}}$

Magnetic induction, $I = \dfrac{M}{T} =$ moment of momentum per

unit area.

Coefficient of induction (self or mutual), $\dfrac{I}{C} = \dfrac{M}{L^2} =$ inertia per

unit area.

This is, or may be, an improvement on the rough practical system which assumes as of no dimensions sometimes K, and sometimes μ, according as one is dealing with electrostatics or with magnetism; but very likely it is only a stepping-stone. Prof. Fitzgerald has recently suggested that, regarding everything from the strictly kinematic and etherial point of view, both K and μ may be a *slowness* of the vorticity; and by that assumption also everything becomes simple and of unique dimensions. Whatever of this turns out true, it is not to be supposed that we can long go on with two distinct systems of units, the electrostatic and the electromagnetic, and two distinct sets of dimensions for the same quantities; knowing as we do that neither set can by any reasonable chance turn out to be the right one.

NEWTON'S GUESSES CONCERNING THE ETHER.

(*q*) Newton's queries at the end of his "Opticks" finish in the early editions with Query 16, and I have found it difficult to come across the later queries except in Latin. I therefore here copy such portions of these queries as have an obvious bearing on our present subject, in order to make them more easy of reference.

"*Qu.* 17. If a Stone be thrown into stagnating Water, the Waves excited thereby continue some time to arise in the place where the Stone fell into the Water, and are propagated from thence in concentrick Circles upon the Surface of the Water to great distances. And the Vibrations or Tremors excited in the Air by percussion, continue a little time to move from the place of percussion in concentrick Spheres to great distances. And in like manner, when a Ray of Light falls upon the Surface of any pellucid Body, and is there refracted or reflected, may not Waves of Vibrations or Tremors be thereby excited in the refracting or reflecting Medium at the point of Incidence ... ?"

"*Qu.* 18. If in two large tall cylindrical Vessels of Glass inverted, two little Thermometers be suspended so as not to touch the Vessels, and the Air be drawn out of one of these Vessels, and these Vessels thus prepared be carried out of a cold place into a warm one ; the Thermometer *in vacuo* will grow warm as much and almost as soon as the Thermometer which is not *in vacuo*. And when the Vessels are carried back into the cold place, the Thermometer *in vacuo* will grow cold almost as soon as the other Thermometer. Is not the Heat of the warm Room conveyed through the Vacuum by the Vibrations of a much subtiler Medium than Air, which after the Air was drawn out remained in the Vacuum ? And is not this Medium the same with that Medium by which Light is refracted and reflected, and by whose Vibrations Light communicates Heat to Bodies,[1] and is put into Fits of easy Reflexion

---

[1] Note the precision and propriety of this phrase : far superior to most of the writing on the subject of absorption of radiation during the present century. It could only be improved by substituting *generates in* for "communicates to," in accordance with the modern kinetic theory of heat.

and easy Transmission? And do not the Vibrations of this Medium in hot Bodies contribute to the intenseness and duration of their Heat? And do not hot Bodies communicate their Heat to contiguous cold ones, by the Vibrations of this Medium propagated from them into the cold ones? And is not this Medium exceedingly more rare and subtile than the Air, and exceedingly more elastick and active? And doth it not readily pervade all bodies? And is it not (by its elastick force) expanded through all the Heavens?"

" *Qu.* 19. Doth not the Refraction of Light proceed from the different density of this Ætherial Medium in different places, the Light receding always from the denser parts of the Medium? And is not the density thereof greater in free and open Space void of Air and other grosser Bodies, than within the Pores of Water, Glass, Crystal, Gems, and other compact Bodies?"[1] . . .

" *Qu.* 21. Is not this medium much rarer in the denser Bodies of the Sun, Stars, Planets, and Comets, than in the empty celestial Spaces between them? And in passing from them to great distances, doth it not grow denser and denser perpetually, and thereby cause the gravity of those great Bodies towards one another, and of their parts towards the Bodies; every body endeavouring to go from the denser parts of the Medium towards the rarer? For if this Medium be rarer within the Sun's Body than at its surface, and rarer there than at the hundredth part of an Inch from its Body, and rarer there than at the fiftieth of an Inch from its Body,[2] and rarer there than at

---

[1] In Newton's opinion light travelled quicker in gross matter than in space, and hence it is that he inverts our Fresnel-derived views. He continues the same inversion in his query concerning gravitation, here next following.

[2] It was his experiments in diffraction which made him think of this gradual change in the properties of ether as one recedes from a body. A few years ago such gradual changes would have seemed to us quite unlikely; but the most recent experiments of Michelson shake all preconceived opinions.

the Orb of *Saturn*, I see no reason why the Increase of density should stop anywhere, and not rather be continued through all distances from the Sun to *Saturn*, and beyond. And though this Increase of density may at great distances be exceeding slow, yet if the elastick force[1] of the medium be exceeding great, it may suffice to impel Bodies from the denser parts of the Medium towards the rarer, with all that power which we call Gravity. And that the elastick force of the Medium is exceeding great, may be gathered from the swiftness of its Vibrations. Sounds move about 1140 English Feet in a second Minute of Time, and in seven or eight Minutes of Time they move about one hundred English Miles. Light moves from the Sun to us in about seven or eight Minutes of Time, which distance is about 70,000,000 English Miles, supposing the horizontal Parallax of the Sun to be about 12″. And the Vibrations or Pulses of this Medium, that they may cause the alternate Fits of easy Transmission and easy Reflexion, must be swifter than Light, and by consequence above 700,000 times swifter than Sounds. And therefore the elastick force of this Medium, in proportion to its density, must be above 700,000 × 700,000 (that is, above 490,000,000,000) times greater than the elastick force of Air is in proportion to its density. For the Velocities of the Pulses of Elastick Mediums are in a sub-duplicate Ratio of the Elasticities and the Rarities of the Mediums taken together." . . .

" *Qu.* 22. May not Planets and Comets, and all gross Bodies, perform their motions more freely, and with less resistance in this Ætherial Medium than in any Fluid, which fills all Space adequately without leaving any Pores, and by consequence is much denser than Quick-silver and Gold? And may not its resistance be so small as to be inconsiderable? For instance ;

---

[1] Meaning what we call the pressure. This is, of course, pursuing the analogy of sound waves, and does not accord with our present knowledge.

if this Æther (for so I will call it¹) should be supposed 700,000
times more elastick than our Air, and above 700,000 times more
rare ; its resistance would be above 600,000,000 times less than
that of Water. And so small a resistance would scarce make
a sensible alteration in the Motions of the Planets in ten
thousand Years. If any one would ask me how a Medium can
be so rare, let him tell me how the Air in the upper parts of the
Atmosphere can be above an hundred thousand times rarer
than Gold. Let him also tell me how an electrick Body can by
Friction emit an Exhalation so rare and subtile, and yet so
potent, as by its Emission to cause no sensible Diminution of
the weight of the electrick Body, and to be expanded through
a Sphere whose Diameter is above two Feet, and yet to be
able to agitate and carry up Leaf Copper, or Leaf Gold, at the
distance of above a Foot from the electrick Body? And how
the Effluvia of a Magnet can be so rare and subtile, as to pass
through a Plate of Glass without any Resistance or Diminution
of their Force, and yet so potent as to turn a magnetick Needle
beyond the Glass?"

---

¹ The interest of these extracts lies largely in their belonging to the
very early days of the conception of an ether, and in their remarkable
insight into many things, though in detail they often do not completely
accord with present knowledge.

INDEX.

E E

INDEX.

A.

E E

Oscillatory discharge, 41, 42, 94,
227—233, 360—382, 396
analysed by mirror, 377
Outstanding problems, 297—303
Overflow of jar, 374

P.

PATHS of energy, 95—97
Peltier effect, 115
Penetrability of ether, 19
Perfect conductor, 165,167,183,184,
195, 261, 274
Period of oscillation, 230, 362—370
Permanent magnetism, 158
magnetism, universal, 283
Permeability, 156, 157, 185, 206,
387
not constant, 283
real value of, 235
Perpetual motion, 135
Phosphorescence, 250, 259, 268
Photophone, 325
Pitch, index of refraction of, 306
prism, 306
Plane of polarization rotated, model
of, 321
Point whirligig, 168
wind, 300
Polarization, electrolytic, 109
of electric radiation, 304
Pole, acted on by current, 134
Potential, 28, 61
of atoms, 77
of isolated metals, 112
of pole on circuit, 393
uniform in conductors, 20
Poynting, 16, 95, 96, 98, 105, 242
Pressure and dielectric-strength, 126
Preston, Tolver, 336
Principia, 318
Problems outstanding, 297
Production of electricity, 313
Projectile method of communication,
335—337
Pyro electricity, 120

Q.

QUANTIVALENCE, 75
Quincke, 282

R.

RADIATION and heat, 66
electric, speed of, 234—256
encountering conductor, 271—
275
exciting currents, 304
loss of energy, 249
maintenance of, 249, 250
mechanism of, 260—275
polarized, 305
production of, 266, 303
reflected, 304
refracted, 306
speed, modes of observing,
236—256
waste, 258, 259
Rails and slider, 207
Range of light waves, 367
Rate of oscillation, 42
Ratio of units, 245
Rayleigh, Lord, 259, 365, 396
Reflected light, amount of, 270
Reflexion, 269—275
by magnetic medium, 279
concentration of light by, 262
metallic, 262
model of, 271—274
Refraction of electric waves, 270,
306
index and spec. ind. cap., 250—
256
Reluctance, 155, 389
Residual charge, 39
Resistance, 32, 69, 70
and impedance, 398
of commutated circuit, 395
to alternating currents, 397
magnetic, *see* Reluctance
Resonance, electric, 304, 374
Retentivity, 159, 161
Return circuit, 28

THE END.

RICHARD CLAY AND SONS, LIMITED, LONDON AND BUNGAY.